Advanced Robot Systems

by

Mark J. Robillard

Howard W. Sams & Co., Inc.
4300 WEST 62ND ST. INDIANAPOLIS, INDIANA 46268 USA

Copyright © 1984 by Mark J. Robillard

FIRST EDITION
FIRST PRINTING — 1984

All rights reserved. No part of this book shall be reproduced, stored in a retrieval system, or transmitted by any means, electronic, mechanical, photocopying, recording, or otherwise, without written permission from the publisher. No patent liability is assumed with respect to the use of the information contained herein. While every precaution has been taken in the preparation of this book, the publisher assumes no responsibility for errors or omissions. Neither is any liability assumed for damages resulting from the use of the information contained herein.

International Standard Book Number: 0-672-22166-7
Library of Congress Catalog Card Number: 84-50184

Edited by: *Pryor Associates*

Printed in the United States of America.

Advanced Robot Systems

Mark J. Robillard has been actively involved in microcomputing for over ten years. He is currently involved in the research and development of advanced color graphics systems and voice recognition peripheral design. He writes for *Robotics Age* magazine and *Microcomputing* magazine and has written *Hero 1: Advanced Programming and Interfacing* and *Microprocessor Based Robotics,* both published by Howard W. Sams & Co., Inc.

Preface

In many technological fields, advances are made every day. Corporate product development programs advance the state of the art by their very nature. Everybody is striving to build yet a better mousetrap. The engineers, technicians, and scientists involved in these endeavors advance their state of learning. It would seem that to advance is the intended natural way.

This book too, being titled *Advanced Robot Systems,* promises to show a chronicle of advancement. In fact, it does this, however, not as advancement in the field of robotics, but rather, as a more advanced volume of information when compared to its previous volume *Microprocessor Based Robotics.* It is true, though, that the information contained here has certain advanced characteristics.

This is the first technical book that stresses the systems approach to the design of robots. Throughout these six chapters, you will find in-depth examples of robot systems hardware and software. At this time, let me summarize each of the chapters to give you a better feel for how all this fits together.

To start off, Chapter 1 leads into a discussion of the design of wheeled "rover" robots. Presently, this type of locomotion is the most popular, given the technological problems involved with other types of motion, e.g., walking. Throughout the hardware discussions, you will be presented a philosophy with which to design by. This constitutes the basis for all systems discussions throughout the book.

Three rover applications are presented. The *automated guideway vehicle (AGV),* used on today's factory floors, the mailroom robot that traverses our office building's hallways delivering interoffice memos and mail, and tomorrow's security robots are each studied in depth. Their hardware requirements and several approaches to satisfying these form the bulk of the information presented in this chapter.

Chapter 2 complements the work presented in the first chapter by showing the software involved in rover systems. Starting off, there is a general discussion of the age-old problem dealing with where the software-hardware split should occur and where it is most cost effective. Several programming examples written in standard BASIC language are given. In instances where machine language would be more appropriate, detailed flowcharts are given on every operation.

Chapter 3 breaks away from the rover mold to deal with the aspects of designing hardware-based control systems for manipulators. In particular, several LSI stepper-motor controller ICs are discussed. The design of an XY table manipulator is presented with applications in the pick-and-place assembly field.

Chapter 4 is probably the most complex; in-depth presentation is given of simulation programs for robot manipulators. Two complete BASIC programs are designed and discussed in their entirety. The first deals with the control of the stepper-motor ICs discussed in Chapter 3. This program allows a personal computer, suitably equipped with a parallel interface (details given), to control the actions of the XY table presented earlier. This control is through the use of English language commands entered through a keyboard or obtained from another computer through an RS232 link.

The second program presented is a full graphics simulation of both the XY table and the control structure presented earlier. This program is for those of you who do not wish to invest the time and effort in actually building robot hardware.

Chapter 5 deals with the popular subject of personal robots. In particular, it deals with the hardware approaches used in both the Heath, HERO 1, and the RB Robot, RB5X. Emphasis, in this chapter, is put on distributed approaches to hardware design.

A completely new personal robot system is discussed and designed in these pages. Here, you will find schematics of ultrasonic rangers, motion detectors, and full robot CPUs.

The last chapter (Chapter 6) ties together all the above under the guise of personal robot software. Actually, many of the concepts presented in all the chapters are reviewed and interwoven to produce a new programming language: ROBOL. Also discussed is the distribution of robot software, adding artificial intelligence and auto adaptive learning techniques.

This volume does not intend to present all that is known about robot systems. Nor does it present all that there is to know; however, it begins by defining the problems we face and shows some creative approaches to solving them.

MARK J. ROBILLARD

ACKNOWLEDGMENTS

I wish to thank the following for their support in the production of this volume and for providing necessary information on that wondrous HERO 1:

Jim Wilson — Heath Co.
Jim Lytle — Heath Co.
Doug Bonham — Heath Co.

and Joe Lombardo
for coming up with his own advanced ideas.

DEDICATION

I would like to dedicate this volume to the following people who have supported me throughout the years.

My mother, Rosemary
My father, Joe
Julia Healey (a supporter from day one)

My special dedication has to go, once again, to Angie, my wife. Thanks go to her for all her work and her many great ideas.

Contents

CHAPTER 1

MOVING PLATFORM MECHANICS 9
Material-Handling Systems—Security Systems and Mailpersons—Motion System's Design—Experiment 1: M1 Remote, Mobile Platform—Experiment 2: M2 Intelligent Mobile Platform

CHAPTER 2

SOFTWARE FOR ROVER SYSTEMS 47
Clock Generation—Control by Planning—Motor Control via Software—Adding a Steering Wheel—Speed Control—Mailmobile Software—Distributed Software—M1 Software Considerations—M2 Software Description—Robot Rovers-Whom Do They Serve?

CHAPTER 3

MANIPULATOR SYSTEMS HARDWARE 85
The Production Environment—Final Assembly Workstation Design

CHAPTER 4

SOFTWARE FOR MANIPULATORS 101
Command Language Design—XY Table Simulation

CHAPTER 5

PERSONAL ROBOT HARDWARE 145
Personal Robotics—HERO 1 Circuit Description—HERO 1 Senses—Advanced Hardware

CHAPTER 6

PERSONAL ROBOTS: SOFTWARE, APPLICATIONS, ADVANCED TECHNIQUES 181
Robot Software—Software Distribution—HERO Executive System—Learn Mode—Advanced Robot Language—Artificial Intelligence—Conclusion

INDEX 209

Chapter 1

Moving Platform Mechanics

When a machine is touted to be adaptable to its environment, it is generally meant that this same machine may be operated under varying conditions. In an automatic assembly environment, adaptability often pertains to a robot's ability to change tool heads or to maneuver to where the work is required. It is this latter ability that this chapter will attempt to shed light on.

In *Microprocessor Based Robotics,* we touched on the mechanics involved in a roving platform. Here the total system necessary to not only make it move but to do so in a precise, planned way is explored. We begin our journey with a discussion of what a roving robot is and how it is being used in industry today.

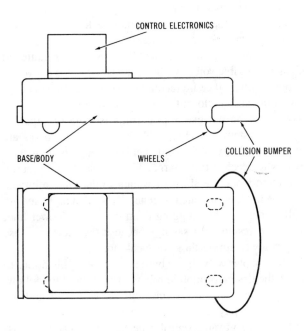

Fig. 1-1. Rendering of AGV rover vehicle.

MATERIAL-HANDLING SYSTEMS

Automated factories are seemingly springing up all over. Their claim to fame is that their output is a steady flow of finished goods 24 hours a day. Although many of today's factories are not making use of all 24 hours, significant savings in terms of raw materials and time have been realized.

Why a savings in materials? That's where we begin our story of the rover robot. It is these machines that are being utilized to carry raw materials to various workstations within a factory in a timely, efficient manner. Once there, they can be used to transport finished goods away from the site, therefore, effecting an ideal work flow. In this way they bring only the materials required at that particular moment in time. There is no need to stock a few spares of parts at the workstation in the instance that the material handler may be late. This cuts down on in-process inventory, which lightens the load on both materials and cost.

The workstations mentioned may be occupied with assembly workers or preprogrammed robot manipulators that mimic the job of the human assembler. The material transferred may be any type or shape. You will often find that in extreme cases where the particular material handled is of an odd or unwieldy shape that the rover platform will be fitted with a special tool or holder that positions the piece for easy mounting or unloading.

Fig. 1-1 shows us a rendering of a typical factory floor rover. The large bumper out front is a piece of spring steel fitted with several microswitches to detect a contact that might result in a collision. The bed of the robot is where the raw material or finished goods are placed. This particular drawing does not show any type of fixturing for special-shaped pieces. All control electronics are usually housed in the back command "tower." Many of these rovers are equipped with their own microprocessors that allow them to be programmed once and then they are free to travel on their route through the factory floor.

Not shown on the drawing is the mobility section of the machine. Obviously, some sort of wheeled arrangement is provided. However, if you remember the discussions from the previous volume in this series, there are many ways to effect motion. Most of the roving platforms used by industry have a conventional three- or four-wheel approach. Of course, there are multitudes of variations in the control methods used to make them roll, and those same microprocessors located in the control head perform that duty also.

Other methods of control are common also. There are some that receive radio commands from a master control station located in a fixed area of the factory. A command may be as simple as needing more material at workstation number 3, or it may be as complex as supplying a complete route map.

The rovers discussed thus far may be classified as *automated guideway vehicles (AGVs)*. They are called AGVs because most of the factories employing them have a complex route guideway "track" located somewhere on the surface of the floor. I hesitate to use the word "track" because more often than not there are no physical rails or even obvious paths that these machines seem to be confined to. They follow a buried wire in the floor where a low-frequency radio signal is transmitted. On-board guidance systems are able to detect the presence of this signal and to assure that they are centered over the guidepath. The wire is usually buried at least a quarter inch below the cement flooring to avoid becoming an obstacle. A local transmitter radios control information to the AGV as to which carrier frequency to follow (track) and can, therefore, effect a change of frequencies that may be used to switch the vehicle over to an adjoining guideway. Later in this chapter we will go over the operation of several types of invisible tracks and the control systems used to effect them.

Factories are not the only place roving robots are showing up today. They are, however, the most prevalent. As time goes on several other types of factory "helpers" will begin to surface.

SECURITY SYSTEMS AND MAILPERSONS

The subtitle "Security Systems and Mailpersons" may sound like a strange mix of disciplines. Actually, the two classes of activities outlined in it are two other common uses of rover robots. In the security field, several rover *droids* are being evaluated as guard replacements. They are equipped with motion-sensing detectors, infrared heat sensors, circuits that detect the noise of broken glass and other security related concerns. They generally are left to roam the grounds of a facility scanning for intruders. They are constantly in touch with a command station inside a building located close by. Any trouble that is encountered may immediately be reported by radio transmission. Several of these "nightwatchmen" are equipped with high-volume sound beepers that will effectively deafen would be attackers or render them helpless because of the pressure on their ears.

Advances are being made every day. Some day perhaps it will be difficult to detect that Old Charley of the midshift is in actuality a complex electromechanical android. Let's hope, though, that the designers of those systems don't allow themselves to incorporate real weapons among the subsystems.

Nightwatchmen and office-building rovers are becoming more commonplace. Insurance companies and other businesses have been profiting from the likes of the computerized mail carrier. These mailmobiles have been wandering down halls delivering interoffice paperwork for almost five years now. They come in various shapes and sizes, from the slender minibus look-alikes to the more familiar cylindrical R2D2-like physique.

Some have several "eyes" that track an invisible phosphorous trail or guideway that may contain instructions as to which places are stop points along the route. When these machines do stop, they generally sound short beeps to make their presence known to the secretaries in the area. After being loaded or unloaded they are free to wander on their merry way to the next office waystation. The period of time they are stopped at a site is automatically programmed into their control structure yet may be overridden by the use of stop buttons located on the body of the vehicle.

What happens if you are in the way during their trek down the hallway? Most of them employ rather sophisticated collision avoidance detectors that will result in the vehicle stopping several inches before the obstacle. Some even have override electronics that will allow them to inch up to an obstruction to nudge it out of the way. However, these nudges will cease if the object is immovable. These rovers are equipped with a large bumper in front like the AGV to guard against any personal injury.

Even though they seem to be more efficient, they will probably never replace the friendly chit-chat that accompanies the morning mail carrier. Perhaps a rover equipped with an intelligible voice synthesizer? Think of the conversation possibilities! It could recount to you its exciting encounter with the broom closet the previous evening!

Power for both these applications is usually supplied through a system of rechargeable batteries. The robots are brought out of service for some period of time to allow their batteries to reach a full charge. In the case of the security guard there would be shifts of vehicles much like a human fleet. The mailmobiles are not generally used at night and are usually recharged by plugging them into the ac power line, which also results in a savings because the electricity use rates are lower during those off-peak hours.

In this chapter we will use both these rover applications to explore the design of mobile robots. Throughout the following pages, various subsystems will be presented and the design philosophy explained. The following chapter will dwell on the software control aspects of the machines discussed here.

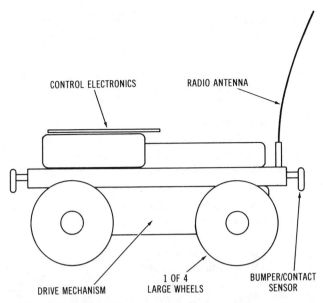

Fig. 1-2. Rendering of M1, remote-control rover.

MOTION SYSTEM'S DESIGN

As it is with anything, thorough planning is essential when designing an electromechanical system. In the building of a house, an architect first must plan the creation by developing a series of drawings called blueprints. In electronic systems a series of design specifications is written. It is here that we begin our exploration of the design of motion systems.

The architecture of the robot is much the same as that of a building. All the critical features and components are listed. In electronic systems the block diagram serves to show, at a glance, all the main parts of the circuitry. Before this diagram can be attempted, however, many things must be determined.

In an intelligent rover some consideration must be made as to whether or not it will require a microprocessor brain. Then, any on-board storage (memory) requirements are hashed out. Finally, the various input and output mechanisms are explored. Sound familiar? It should because that is the general way in which any microprocessor system is planned. But, is it the right way to plan a robot?

Looking at the overall plan for a roving robot design would help with its development. The following list depicts some of the characteristics of a security rover. Of course, the drive mechanism is included and the various electronic surveillance sensors are listed:

IMPORTANT CHARACTERISTICS OF A SECURITY ROBOT

- Three-Wheel Drive Mechanism
- Motion Sensor
- Infrared Heat Sensor
- Voice Output to Warn Intruder
- Radio Link to Base
- On-Board Intelligence
- Rechargeable Power Source

Notice where the mention of the brainpower comes? It is generally thought of as secondary. The drive is tantamount because it is the most impressive of its capabilities. The sensors are what makes it do its thing. The brain is simply there to tie those together.

After this assessment of the system, shouldn't we start with the I/O and slowly migrate to the processor? Not yet! Remember, as robot "architects," it is our job to first draw up specific specifications for each sub-unit within the entire system.

Here is where we go separate ways. The first application studied will be a remote-controlled rover. This system will be basically brainless (at least there will be no on-board brains). This type of system may be used as a police bomb disposal unit or even a no-thrills security drone that may be operated from a central guard station.

Fig. 1-2 depicts a typical system. I will call this M1 for Mobile 1. M1 has several features. The first of which is its four-wheel drive base. The rubber tires are somewhat deflated to ensure increased traction. Its top speed may be two feet per second, and it can vary this speed, proportionally, down to a slow crawl. The robot itself only measures 15 inches in length and weighs about 4 pounds. The weight is mainly attributed to its six C-cell batteries that power the base. These are rechargeable nickel-cadmium cells (nicads) that are mounted in a compartment in the undercarriage. The wheels are driven by gear motor systems that receive their commands as to direction and speed via RF carriers. The two front wheels steer 15 degrees in either direction proportionally, much like the speed control. Both controls come from the same radio receiver.

Feedback from M1 is in the form of a front-mounted collision bumper. When activated, a separate radio frequency is transmitted from the vehicle indicating a collision. This signal is sufficiently different from the received motion signals that no interference is possible. Both transmitter and receiver are mounted inside the body of M1. There are no other inputs or features.

What good does this remote-controlled base do us? You'll see later, just hang on.

The next application to be introduced is fondly called M2 (hard to come up with that name). In contrast, M2 has brains. In fact, there is very little that M2 does without the use of at least one microprocessor. Yes, I am talking about a multi-processor vehicle. This unit has only two wheels. They work on the principles outlined in the first volume when I described Milton Bradley's Big Trak®*. In fact, one of these bases is used in this experiment. A separate slave microprocessor is in control of these wheels. Its language is strictly "which direction" and "how far." It off-loads the main microprocessor of

---

*Big Trak is a registered trademark of Milton Bradley.

the humdrum chores of motion. There is another microprocessor dedicated to the ultrasonic ranging system. In this application, M2 can maneuver in rooms without getting lost or bumping into walls.

M2, like M1, is also powered by batteries located in the base. However, it also requires several other ones throughout the superstructure of the robot. The reasons behind this will be pointed out soon also. Is this starting to sound like a sales pitch? Actually, it is! I want you to start daydreaming about how these things are interconnected and why. That is what design is all about. We have only speculated on the design of each of these applications. There are no block diagrams as such. You see no schematics, no program listings, no parts lists. It is here that the concept of "imagineering" begins, and to be an effective robot designer you must have a great deal of imagination.

Design Specifications

Now we get back to earth. It's time to get to work listing actual specifications for each application. The place to start is to define a list of functional duties. For our security robot the following list spells out my desires. Notice that this list is quite a bit more detailed than the previous one. Here is the place to get your wishes in. You may find later that the duty you would like to perform is either too costly, too difficult, or even impossible.

DETAILED REQUIREMENTS OF A SECURITY ROBOT

- Drive Mechanism
    Three wheels
    Each wheel powered
    Special traction tires
    Proportional steering
- Motion Sensor
    Ultrasonic ranger for distance to object
- Infrared Heat Sensor
    Determination between man and other objects
    Fire detection
- Voice Output
    Highly intelligible
    Canned phrases
- Radio Link to Base
    High reliability
    Voice and data
- On-Board Intelligence
    Low power consumption
    Performs simple, logical deductions
    CMOS nonvolatile memory
    Field-replaceable snap-in package
- Rechargeable Power Source
    Gel-cell battery supply
    Charger built into rover

The following two lists outline similar desires and functional duties pertaining to the mailmobile and the factory AGV-type applications.

DETAILED REQUIREMENTS OF A MAIL-HANDLING ROBOT

- Drive Mechanism
    Four wheels
    Two wheel steering
- Optical Path Sensor
    Three-detector system
    Follows path by straddling edges
    Middle sensor detects stop points
- Ultrasonic Collision Sensor
    Detects 2 feet in front
    Slows vehicle to crawl
- Mechanical Bumper
    Stops vehicle when activated
- Travel Override Switch Bar
    Halts vehicle from resuming route
- Stop Point Annunciator
    Beeper to signal stop point
- On-Board Intelligence
    CMOS single-chip processor
- Rechargeable Power Source
    Gel-cell battery
    Built-in recharger

DETAILED REQUIREMENTS OF AN AUTOMATED GUIDEWAY VEHICLE

- Drive Mechanism
    Four wheel
    Two steerable wheels
    Proportional steering
- Radio Guidewire Sensor
    Two-coil type sensor and control
- Mechanical Safety Bumper
    Stops vehicle immediately when activated
- Automatic Warning Light and Bell
    Lights and sounds when vehicle is in motion
- Radio Command Link
    Separates command frequency link to controller
- On-Board Intelligence
    CMOS controller links to sensors and radio
- Rechargeable Power Source
    Lead-acid car battery

Where does the radio unit fit in? Hold your horses, that's a way down the road!

With the basic functions defined, it is time to put quantitative amounts on the wishes that will turn them into specifications. Determine how much that arm can lift, how fast variable speed is, and what type of microprocessor you have in mind. Start defining the size and shape of the robot. Its

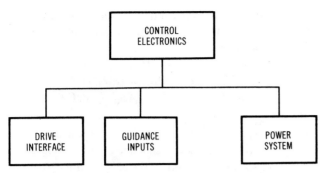

Fig. 1-3. Block diagram of a mail-carrying rover system.

weight is specified here and any power requirements. List them all. It will make the design process easier.

After this is complete you have about enough information to build a block diagram. This document shows the functional interconnections between sub-units in an electronic system. Normally the system block diagram is void of details. It may be as simple as showing a microprocessor, I/O, memory, and a motor interface. Don't make it too general, for then it would serve no purpose. I have laid out a general system diagram of the mail robot shown in Fig. 1-3.

Don't worry about my going through the robots too fast because I intend to go into the excruciating details of their design later.

As you can see in the figure, all the main functions are spelled out. There is no mention, however, of which type microprocessor or even details on any of the I/O circuits. Yet this is a specific diagram of a functioning mail-carrying robot. From here, a detailed circuit block diagram or block diagrams are done. The reason for the mention of more than one is that microprocessor brains for robots may require their own tiny system of various functions, as well as some of the I/O circuits. You will see later that the design of an intelligent motor drive interface can become as complex as a personal computer.

Drive Considerations

It's time to build our block diagrams for real. Using the hypothetical one shown in Fig. 1-3, the first block to be defined will be the drive interface. As I said before, this is the most obvious feature of a roving robot. It is also the one interface that must work before the rest of the system is operational. When considering a drive system, several factors are involved. I am not going to go over them in great detail because they are covered in Chapter 3 of Volume I. I will, however, revisit the principles of a two-wheel drive system.

Obviously, depending on the application, the drive mechanism of a particular rover may be entirely different. In our AGV application, a three- or four-wheel, conventional steering mechanism type system would serve the need. In the mail carrier this may also be true. The security robot may demand more control over rough terrain though. In this application, a treadlike arrangement might prove essential.

In the specific design experiments described as part of this chapter, I chose both a four-wheel and a two-wheel mechanism. The two-wheel type may sufficiently simulate that of a tread. Both are not industrial vehicles but scaled down working models that incorporate the same electronic systems design of the larger more powerful types in use today.

Let's get into the drive properties of both now, in order to complete the block. Fig. 1-4 depicts both systems used here. In Fig. 1-4B the four-wheel system uses two stationary wheels that are powered by a dc permanent magnet motor. Changing the direction of current through the motor effectively reverses it. The two front wheels are movable. They are allowed to pivot about a parallel axis 15 degrees in each direction. This will serve to guide the vehicle to the left or the right during motion.

These two steering wheels are also powered to turn in the same direction as the rear ones. This gives a four-wheel drive effect. It increases drive traction and allows the vehicle to climb upgrades and traverse rough terrain.

The two-wheel drive mechanism uses a separate motor for each individual wheel. There is no common axle between wheels. Forward motion is accomplished (see Fig. 1-5) when both wheels are rotating in the same direction. Turning happens when one or the other motor is reversed. Refer back to this drawing throughout the discussions of M2's drive control circuitry.

A three-wheeled system would act in much the same way

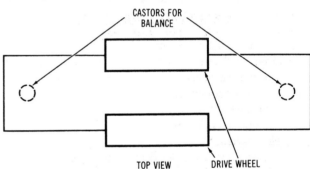

(A) Two-wheeled drive.

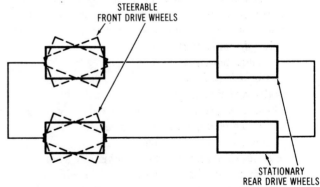

(B) Four-wheeled drive.

Fig. 1-4. Mechanics of two drive systems.

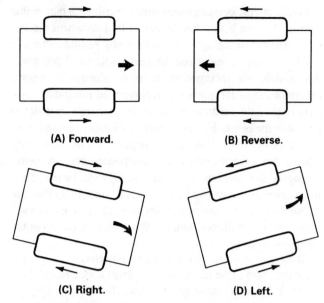

Fig. 1-5. Action of two-wheeled drive under varying wheel direction combinations.

as the four-wheeled design presented. It is not necessary, of course, to power that third wheel. Just use it for guiding the vehicle to the right or the left. Other forms of drive mechanisms exist. They are only limited by the imagination.

Pick the drive system that suits your application best and put it into the drive block in the diagram. Let's go on to the input section of the system.

Input devices for rovers can vary from bumpers that detect collisions to manual push-button switches used to delay the mail carrier a little longer. Basically, there exists a myriad of sensors that get attached to this input block. Also, here is where any thought toward a sophisticated guidance system begins.

Rover Guidance

Guidance of a moving vehicle is a very difficult problem. You're probably going to feel that this part of Chapter 1 will go on for three chapters. In actuality, the discussion will be broken up between this and the next chapter. Only hardware issues are covered in this chapter. The software used to guide vehicles is presented in the next chapter.

Let's assume that, for our purposes, guidance is the act of assuring that a vehicle reasonably follows a preset course or route. This course may be physically marked or constrained, or it may be completely invisible. The vehicle will, at some point, be given the possibility of finding this route, and it will have the ability to physically follow it. There are a number of ways to mark a route. Let's look at some of them.

The first is to simply draw, with a contrasting color, the route to be followed on the ground or flooring that passes beneath the vehicle. Another variation to this could be to paint an invisible line that only is detectable under certain types of lighting. This is similar to the ink that is used in some amusement establishments to mark your hand. They can tell if you have already paid for your ticket by shining a portable ultraviolet lamp over the area. There are several chemical solutions to accomplish this. The invisibility is used mainly where the route markings would appear unsightly as in an office environment.

If you're roving around a factory floor, there are even other alternatives. Standard single and double tracks are quite acceptable. The railroad train guidance system is obviously the most simple, as long as the train doesn't exceed a set speed limit. This might cause the momentum of the vehicle to push it off the guide track. If your workers find that those tracks get in the way, there are invisible tracks that can be implanted beneath the floor.

As discussed briefly before, an RF carrier is transmitted over these invisible tracks and is picked up by small receivers located in the vehicle. Depending on the strength of the signal, the vehicle will steer proportionately to the right or left. Through this system, a very accurate path control mechanism can be accomplished.

There are a few other methods of guidance being explored or used in industry. One is to use optical rangefinders located on the rover. Special reflectors are located in various places around the plant. Through triangulation the robot can find its position and make any corrections. A similar system uses three lasers located in corners of the plant. On the rover is a sophisticated rangefinder mechanism that detects the presence of these signals and computes its position by determining the phase relationship between itself and the beacons. We will begin to investigate the design of several of these guidance systems here but we will leave out the more exotic.

Optical Track Guidance — The method used by standard railroad cars for guidance is obviously the simplest implementation of keeping a moving vehicle on a predetermined path. Outside of this purely mechanical method, few approaches are as easy to design as that of the optical track.

As mentioned before, the optical track system is based on the detection of a painted line that is applied to the floor surface beneath the path of the moving vehicle. In the case where invisible paint is used, a special lamp will be used to illuminate and, therefore, detect the presence of the path.

There are several approaches that can be used to both detect and follow this painted line. All methods use the reflection of a beam of light off the surface of the path or adjacent to it. Fig. 1-6 illustrates the basic principles of this approach. A beam of light is transmitted from either an incandescent or solid-state lamp source in an angular direction toward the surface directly underneath the vehicle. The laws of light reflection show us that some portion of this beam will bounce off the surface and be deflected toward a similar yet opposite angle. It is at this angle of deflection that a mechanism is employed to capture the energy of this reflected beam. This

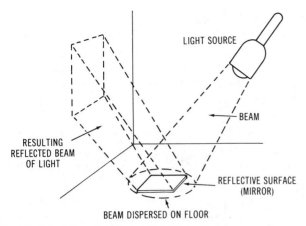

Fig. 1-6. Principles of reflected-light detection.

device may be one of several types of photocells, phototransistors, photodiodes, or even the somewhat outdated photomultiplier tube.

Because of the proximity between the floor surface and the emitter-detector pair, there will always be some amount of reflected light falling on the detector. The normal coating that is applied to the floor will reflect a given amount. If the painted line is reflective in nature (as in a metallic strip), then more light will be detected when the guidepath is encountered. Conversely, if the path is painted with flat black paint, then a lesser amount of light will outline the guideway.

With these basic principles in mind, it is now possible to venture further into the control method of staying on the guideway once it's been detected. The first approach to study is the single-head method. In Fig. 1-7 a simple mechanical and electrical design is outlined. A vehicle is fitted with a guideway emitter detector, which is mounted in the front center of the vehicle. The amount of light reflected from the roadway is converted into a proportional voltage by a phototransistor (detector). This voltage is then applied to one input of a two-input voltage comparator; the other input is connected to a manually variable voltage source. When the vehicle is initially turned on, this manual level must be set up to provide a trip point for the comparator. With the detector not looking at the guidepath, the variable source is adjusted until the voltage from the detector equals that of the manual source (reference). At this time, the output voltage comparator will switch to an active state. Any significant change in the light energy detected will cause the comparator to change state indicating a not-equal-to condition between the two voltage sources. In this way, the guidepath is electronically detected. From here on, the control of a vehicle is relatively simple.

Fig. 1-8 shows a simple circuit that is used to control a two-wheeled base using the single-detector method. Following the drawing, we see that as long as the comparator output is not active (on guidepath) the wheels are connected for a steady forward motion. When the guidepath is lost, as in the case of a turn, what happens? According to this schematic, the vehicle backs up! Obviously, this is not an adequate approach. How will the control electronics know which direction to turn so that an encounter may take place once again with the path?

This is where a second detector comes in. It is possible, with a smarter electronics package, to utilize the single-detector method for guidance. However, at this time, let's stick to the simpler methods and leave the more complex for the next chapter when more program-oriented control is discussed.

When using two detectors, there are two methods that may be employed to mount them. One is to fix them so that their light beams will straddle the guidepath. The circuit in Fig. 1-9 shows this type of implementation. As long as the path is not detected, the vehicle goes forward. When the right detector sees the path, it turns the vehicle toward the right to straighten it out. The same is true for the left except, of course, the vehicle will turn to the left. This side-to-side travel may be reduced if the detectors are spaced to where they are very close to the path.

The other method of using two detectors involves having them on either edge of the path. Fig. 1-10 depicts the circuit, and the actions the vehicle takes when one or the other loses the path. As you can see this implementation is very similar to the single-detector method except it is now possible to detect which direction the vehicle is traveling in when a loss of guidepath is discovered.

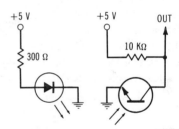

(A) Emitter-detector schematic.

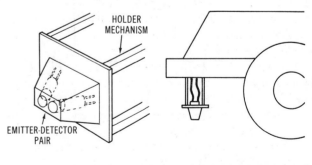

(B) Holder mechanism. **(C) Vehicle with sensor mounted on front.**

Fig. 1-7. Mechanical details of a simple one-head optical guidance detector.

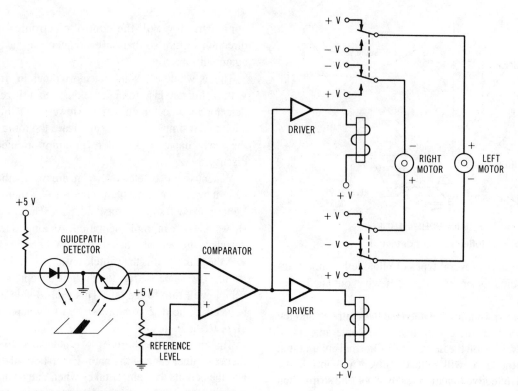

Fig. 1-8. Schematic diagram to support the design shown in Fig. 1-7.

In my experimentation I have found the latter of these two to be the most reliable. Some of the problems encountered with the straddle method were the fact that the change in direction of the motors, as a response to a detector change, was, in some cases, not sustained long enough to effect an actual physical change. In those cases, the vehicle would perform a slight jerk; however, once the detector cleared the path, it assumed that it was successfully straddling it again.

Another problem in all methods has been motor noise-tripping circuitry that enabled the relays at odd times. This can be solved completely by introducing a separate power source for the motors.

All the previously mentioned methods attempted to keep the vehicle straddling or on a painted guidepath. A reverse approach to this could be designed where the guideway is actually composed of two painted lines resembling a road with curbways! Fig. 1-11 spells out the particulars. In this

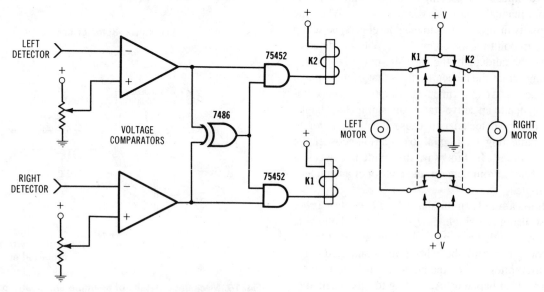

Fig. 1-9. Schematic diagram of a two-detector guideway system.

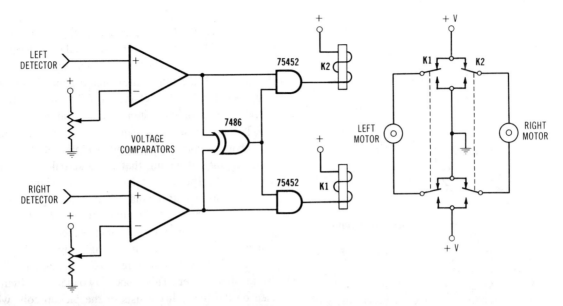

Fig. 1-10. Schematic diagram of a different method of implementing Fig. 1-9.

system the control electronics assumes that as long as both detectors see no guidefence, all is okay. When the right detector encounters the fence, it will switch directions of the left wheel to pull it toward the left, therefore, guiding it back on the path.

As you can see, there are several methods to guide a moving vehicle using light and a painted guidepath. In the case of the mail carrier, at certain junctions along the path instructions are embedded into the path. These commands tell the vehicle that this is a proper place to stop and announce itself. From that point on an electronic timer, within the workings of the rover, takes over to determine how long to stay there. As mentioned before, several manually operated halt switches may be located around the perimeter of the vehicle through which this time may be extended.

The information coded into the guideway is usually done as a series of short guideway "gaps." These are actually areas of the painted path that are void of paint. A special information gathering detector is mounted in the front center of the vehicle to read the path for these gaps. If you liken the guidepath to a series of painted stripes like those found on the side of a grocery product box, the approach becomes clearer. Although the information provided underneath these vehicles is seldom of that complexity, it could be possible to use the exact or a derivative of the Universal Product Code.

You may have noticed, if you have read the previous volume in this series, that optics plays a big part in the field of robotics. Here, we have just begun to touch on the application of light guidance systems. The painted path is by far the simplest. Others involve triangulation techniques much like

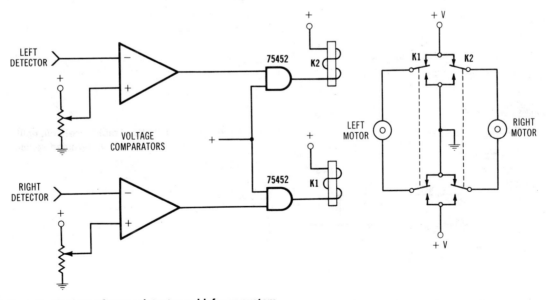

Fig. 1-11. Schematic diagram of a two-detector guidefence system.

17

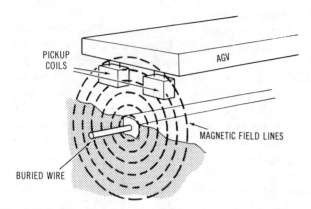

Fig. 1-12. Basic principles of wire-guidance-system operation.

radar. Let's go on to the other popular method of rover guidance used today.

Buried Wire Guidance — You may recall from your high-school physics class that around every current-carrying wire there exists a magnetic energy field. This field, depending on the power of the signal applied to it, may radiate through the air quite a distance. It is on this basic principle that wire guidance systems are based.

A wire or series of wires are arranged in an orderly fashion on the floor of the plant. A signal is then applied to this wire bundle and voilà a problem with the FCC! All joking aside, the field that radiates from the wire is now the guidepath. The moving vehicle must have the ability to detect this energy and determine if it is over the wire. This is done very similarly to that of the optical guidepath. In Fig. 1-12 you can see that a receiver coil mounted on the undercarriage of the vehicle picks up the signal. An amplifier is needed to beef up the strength so that it may be applied to one input of a standard voltage comparator.

As was the case for the previous designs, this comparator has a manual input source that is adjusted to the level received when not directly over the wire. In that case, the two inputs should cancel each other out, which triggers the output to be in the active state. Any change in field strength (like being over the wire) would result in the comparator's output changing. This is then sent to a control system also very similar to the previous approaches. Sounds simple right?

In actuality, detecting that radio signal is probably the hardest task. The air is filled with stray signals, and in a factory environment, the machinery emits a great deal on its own. Therefore, specially tuned circuits must be used to perform this detection. If you want to experiment with a wire-guided system, I offer the following design.

Fig. 1-13 shows a simple receiver schematic (we'll get to the transmitter later). This is one of two identical circuits to be used on the rover. It consists of the pickup coil, which is composed of approximately 200 turns of 26 AWG wire wrapped around a 1- to 2-inch bobbin or similar coil form. The coil is then mounted on the undercarriage so that it is in the same plane as the guidewire. In this case, the guidewire may be considered lying parallel to the floor. Therefore, the coils should be positioned horizontally to the floor as shown in Fig. 1-14.

The coil is acting as the receiver's antenna. The signal from the guidewire is a low-frequency sinewave so the receiver detects a slowly varying signal. The amplifier used here is an extremely sensitive wide-band operational amplifier denoted by the CA3035 (Fig. 1-13). The circuit is tuned to accept the audio range signal by the selection of the coil and the value of the capacitor connected directly across it. Its internal three stages of amplification add up to provide a whopping gain of

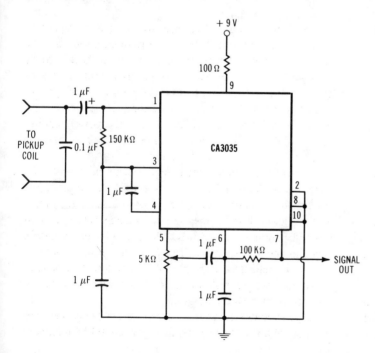

Fig. 1-13. Schematic of radio receiver/amplifier for experimentation with wire-guided vehicles.

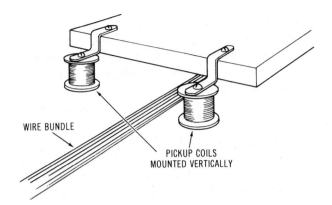

Fig. 1-14. Mechanical details of coil mounting in a wire-guided vehicle.

about 100,000 times that of the original signal. The component selection around the circuit is done to provide very low frequency (such as that of 60-Hz power lines) rejection. In all, this circuit is very sensitive.

The transmitter for this experimental system may be your home stereo. That's right! Now that you've taken over the basement, den, and kitchen with your electronic fooling around, why not ruin the last piece of good entertainment equipment you have left! Actually, you don't even have to take the cover off. The normal speaker connections provide more than enough power for an experimental system.

Simply lay out an oval pattern about 3 to 4 feet in length with about 100 feet of 28 AWG wire over the pattern. This is your guidepath. Connecting this to your stereo will provide a strong transmission signal. The receiver actually becomes the secondary of an air-core transformer as the signal passes from the transmitter loop antenna to the receiver coil. There you have it. Instant radio interference!

The method discussed thus far, regarding signal strengths in the guidewire, is not the only implementation. A recent patent application has been filed on behalf of the National Aeronautics and Space Administration (NASA) that describes a wire-following technique that uses several pickup coils. The phase difference between them determines the location of the wire.

Control Electronics

There are other methods of pinpointing a rover's position. Several ultrasonic and infrared circuits relating to this are outlined in Volume I. From here, let's leave this block of the overall plan and move on to the central control block. Here is where the decision is made to go with either a hard-wired single-function machine or one that is reprogrammable.

Obviously, the trend today leans toward programmability; however, there are still a number of dedicated-function systems. The optical-guidance-control circuits we discussed earlier are a good example of the hard-wired logic I am describing. If you wanted a two-detector vehicle to guide itself many different ways, you would have to wire it for each of those applications.

The big advantage that dedicated logic controllers have is that they require no software development and are generally faster to design. If the job calls for a single function, it is usually more economical to pursue the hard-wired system. There are no real guidelines or rules governing the implementation of a hard-wired system. Just take into account the duty the controller is to perform and perhaps the power system availability. There are hard-wired controllers that are built mainly of relays. In these applications the power consumption of the controller is negligible compared to the power to be supplied to the motors. Older guided vehicles used relay control systems.

As you can imagine, there are several disadvantages to going the dedicated route. The first of which is that should anything change in the application, the logic must be redesigned. In the early computer days, this problem was dealt with by providing large patch panels that were wired using patch cords with phone connectors at each end for each application. They even went as far as to make removable patch panel boards that were prewired and stored on the shelf. Whenever the unit was to be *reprogrammed*, another patch board was plugged in.

This short trip back into history points out yet another disadvantage of the hard-wired controller. The size and resulting weight of such systems can be a disadvantage. Obviously, in systems comprised of relays there is a bulk problem. However, in units designed today of low-power CMOS circuits, it is debatable whether or not they are a hindrance.

Today, even some home ovens are programmable. Microprocessors are evident in practically all areas of control. Their low cost makes them a "shoe-in" for dedicated appliances. If a blender can utilize a microcomputer as its controller, why shouldn't a rover robot.

There are obvious advantages to using a computer to control a robot vehicle. Several modes of operation may be performed. It may be programmed to be adaptable to its environment. To an extent, it can program itself as it goes along. A true intelligence can be designed into the system, and the costs are certainly low enough. There are CMOS microprocessors that effectively reduce the power requirements, and their size on a function-for-function basis is smaller than the hard-wired version.

As mentioned before, the software required to make the controller work may prove to be a drawback. Microprocessor systems are generally more susceptible to noise in the industrial environment and, therefore, more care must be exercised when designing the system. But these disadvantages hardly outweigh the obvious gains that a computer brings to a robot vehicle. In light of this, I have chosen to pursue the course of explaining rover control systems using microprocessor components throughout the design.

Programmable Control — If you're the proud owner of the first volume of this series, then you should be relatively

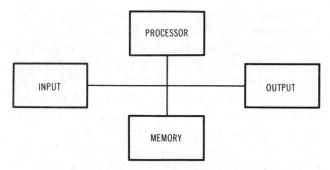

Fig. 1-15. Simplified diagram of the functions of a programmable controller.

familiar with this subject by now. For those of you who have yet to pick up that book, I'll cover the important particulars here, which should tide you over long enough for you to get the first volume.

In robot systems there are two generally accepted implementations of programmable control. The standard microprocessor approach yields a block diagram similar to that of Fig. 1-15. This figure is obviously an oversimplified likeness of what the system would reflect. However, it makes the point that there are four major functions within the system. The input section is comprised of any sensors be they for collision, guidance, etc. Output circuits would control motors for the drive system and any manipulator electromechanical components.

The main microprocessor and its associated memory storage area would coordinate all the rover's activities. In a typical situation, the microprocessor might command the drive system to move forward, then enable a collision avoidance system to detect forward obstacles. This one processor would have the job of physically pulsing an ultrasonic transmitter while detecting sonic reflections off objects. During this process, it must also feed whatever information is necessary to the drive circuits in order to keep the wheels turning.

This may seem a formidable task for one processor, but considering the speed at which microprocessors run and the speed at which the rover is moving, there really is plenty of time to do a lot of things. However, having one processor do all this would require an extensive software executive program. Usually these all-in-one programs are designed for one configuration of rover. If, for instance, you decide to add some functionality to the system, many times the software must be rewritten to accommodate the new addition.

This type of limitation almost resembles the hard-wired controller problems. Remember that, in their case, the problem was somewhat alleviated by providing several plug boards that were prewired to different programs. These boards would be selected, at will, by an operator depending on the particular process or duty the machine was to perform in that instant.

For programmable control there are a number of ways that this feature of operator-selectable programs can be implemented. The first that comes to mind is the removable ROM cartridge so often used in home video games. Using this approach, the rover can be made to adapt to several (if not an unlimited number of) different situations. You could have a pack that programs the system as a mail carrier. Another cartridge would change its personality into a material-handling AGV.

As you can see, in a standard microprocessor architecture system, the possibilities are limitless. The ROM cartridge idea, although sound, does present a problem though. It would be physically necessary for the operator to actually change the program by replacing one cartridge with another. In cases where the vehicle is inaccessible to human operators, this may prove to be a problem.

The way around this deficiency is to have several programs coexistent within the main rover executive. Each one could be called up according to plan automatically. Unfortunately, this way suffers from the fact that, should the hardware functionality change, so goes the executive. Therefore, a self-adapting software system that would automatically configure itself according to the available hardware features would be most desirable. In the last chapter of the book I will go over the design considerations for such a program.

As technology has progressed, the standard microprocessor architecture has somewhat diminished in size to where whole computer systems are now contained on one chip. In Volume I, a tutorial on the design and use of one of those microcomputers is presented. In small, relatively simple rover systems, the use of a single-chip computer will serve to significantly reduce both the cost and the physical size of the controller portion of the design.

In other more complex applications, these microcomputers would serve better to off-load a larger system of several mundane I/O chores. Some perfect examples of this type of delegation of workload would be where the collision avoidance system and all bumpers were to be monitored by one of these *systems on a chip*. In the event of a collision, the main processor in the rover system would simply be notified. There is no need for the system to be involved in the lower-level tasks.

Distributed Control — What is described before is being called *distributed control*. Distributed in that the major control functions are spread around to several mini-controllers.

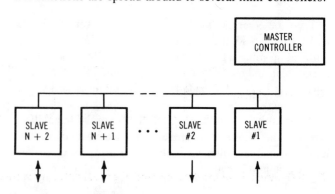

Fig. 1-16. Concept of distributed control.

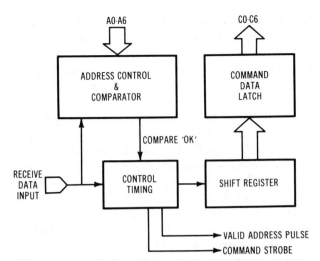

Fig. 1-17. Functional block diagram of the MC14469 receiver section.

In effect, this type of control architecture becomes a sort of communications system where each node is a controllable function. Each function reports back to the main task controller when called upon. Fig. 1-16 illustrates this concept.

The communications path between nodes may be implemented several ways. In a standard microprocessor system this would represent the microprocessor bus structure with its data lines, address lines, and various control signals. You will soon find out that in an industrial electromechanical system such as a rover, routing the microprocessor bus around the frame of the system will only serve to attract electrical noise that will destroy the system's integrity. In these cases, a more noise-immune implementation must be chosen.

Communications links have suffered the same problems over the years. Yet today we effect extremely reliable data highways between systems. The implementation that eventually solved most of the noise problems of the early days was a cross between an alternate hardware system and a new layer of software called *error correction and detection*.

Within rover systems it is possible to apply the knowledge gained in the communications field and, therefore, short circuit most of the systems-type problems that usually crop up in an electromechanical design of this type. Send the data serially and provide for smart communications controllers to carry out the inter-function dialog between nodes.

Recently, there have been advances in single-chip technology that have affected this very application. There are three vendors of integrated circuits today that are providing intelligent controller node components. Let's examine each system for operation and hardware implementation, and where they would fit in the design of a rover system.

Motorola, Inc., MC14469 Addressable Asynchronous Receiver/Transmitter—This device, made available around 1978, was the first higher-level communications part. By higher level I mean that it went one step beyond the duties of simply taking parallel, 8-bit data and shifting it out serially, and then reversing the function. This part incorporated an internal bit-rate timing generator that allowed a lower component count system to be implemented. It also was the first to include device address recognition circuitry. This feature would allow several of these devices to be physically connected to a single communications line. Each device possesses its own unique address, which is the first serial byte of data passed from a master controller. If the address matched, the device would then convert the next serial data byte into a parallel set of bits. If the address did not match, it would simply wait out the next byte for the following next address message.

In effect, this component could serve to process data being sent to a remote distributed function. It only allows data in that is meant for that function by comparing an address code against its internal hard-wired address. Let's examine the

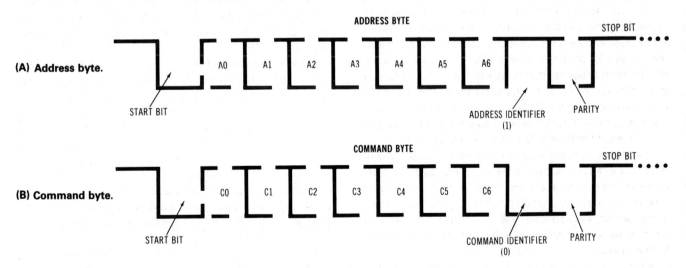

Fig. 1-18 Structure of the data received in an MC14469 application.

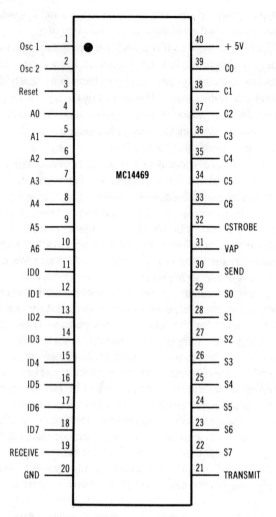

Fig. 1-19. Pin designations of the MC14469.

method with which this selectivity is achieved within the Motorola, Inc., part.

Fig. 1-17 depicts a functional block diagram of the receiver portion of the MC14469. A cursory look will reveal a standard UART-like serial-to-parallel converter. The parity and framing check function is also standard. There are, however, two functions shown that are unique to this part. The address comparator and data comparator and the command output section allow this part to be addressable.

Let's examine the structure of the data this part is expecting to see. For a while refer to Fig. 1-18. This figure shows that 2 bytes of data are usually transmitted from the master controller. These 2 bytes constitute a complete message. The first byte is comprised of a start bit, 7 address bits, an address identifier, a parity bit, and 1 stop bit. These 11-bit times are to be referred to as the address byte.

The receive control circuits, within the MC14469, check the input line, constantly, for the telltale negative transition of a valid start bit. As is the case with standard universal asynchronous receiver transmitters (UARTs), spurious noise glitches are automatically rejected if the transition is not as long as half a normal start bit. In most cases, noise will be quite a bit shorter than that. If, however, you are operating at extremely high bit rates, this kind of noise problem may become significant. As a rule of thumb, keep the bit rate slow enough so that the longest noise pulse width is less than half the normal start bit time.

If a valid start bit is received, the control circuits then begin accumulating the following 7 data bits. Each bit is clocked into the control circuitry at the center of its bit time. The bit following these 7 data bits will tell the part if this is address or control data. In the case of this first received byte, this bit had better be a logic one.

Upon detecting the address identifier bit, the previous 7 data bits are treated as an unsolicited request to verify an address. At that time, those 7 bits are loaded into the address compare circuitry. The receiver then refocuses its attention back on the serial bit stream awaiting the parity bit.

This system is internally set to expect even parity. An even parity indication would be shown if the state of this bit (either high or low) would cause the total number of logic ones in the address word including the address identifier and the parity bit to be an even amount. If even parity is detected, then the address compare circuitry is enabled.

Each MC14469 part includes seven input pins that are used to hard-wire an address to. Seven bits indicate that the maximum number of these parts that may share one communication line equals 128. In a rover system, where each address would represent one node, you will find this number to be more than enough. Getting back to where the part has received the address byte, if the parity checks out okay, this address will be compared to the state of each hard-wired address line. If both address codes match, a pulse will appear on the valid address output pin (Fig. 1-19). This signifies that this particular part has been addressed. It could be used by external circuitry to trigger some event.

In a distributed rover control system each of these nodes may perform some specific control function. In the case of a motor control node, the MC14469 would be used as, perhaps, a remote output port. Several output bits would be required to effect the control of a motorized platform. Fig. 1-20 illustrates a typical node in connection with a motor control function.

In the diagram you can see that there are direction and enable control signals for each of two motors, which totals three possible control output bits. If the system were to incorporate a steerable guide wheel, perhaps more bits would be necessary. This example, however, fits most vehicle applications. You will see where the second byte transmitted to the MC14469 fulfills this control function.

When a valid address has been detected, an internal status latch (valid address latch) is automatically set. This event signals to the control circuitry that a subsequent byte of information may be allowed to enter. This next byte is called the *command byte*.

The command byte is structured similarly to the address

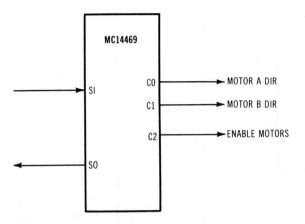

Fig. 1-20. Diagram showing the use of the MC14469 as a motor control remort part.

byte in that there is a start bit, 7 real data bits, 1 bit that signifies that this is a command byte (as opposed to an address byte), even parity indication, and an appropriate stop bit. The eighth data bit that actually indicates which byte is being sent will, in this case, be a logic zero.

What happens when one of these command bytes enters the receiver of the MC14469? Assuming there was a valid address byte before it, the data bits (all 7) will be latched in the command latch. These output bits are available to external circuitry through pins on the package (refer again to Fig. 1-19). In conjunction with the data arriving on these outputs, a command strobe output pulse is generated for use by external circuitry.

Fitting what is just described about the capabilities of this part into the function node diagram of Fig. 1-21, you should begin to see the possibilities. A remote output port separated from the master controller by two wires will greatly add to the flexibility of the rover system architecture. Several output nodes may be connected to these wires in a party-line fashion. Each one responds only to the data being specifically addressed to it.

Communication software on the master controller side is simplified to where the entire output structure resides at one location in its memory or I/O map, and a single output driver can be written to accommodate many varied types of devices.

What about these two wires I've been alluding to in all these paragraphs? What type of transmission path is this? The simplest implementation to connect the MC14469 with is the standard TTL-level signals. The part is expecting to receive an electrical equivalent to this anyway, and the standard UART that acts as the master controller's mouthpiece usually outputs this type of signal also.

What are the drawbacks of using TTL in a rover distributed system command link? Well, before we get into that, let's look at how the MC14469 physically connects to the line. Fig. 1-22 is a schematic representation of the two-wire link to the part. Simple isn't it? I can tell you that on a lab bench with a microcomputer serial output connected to this part's receive input (appropriately interfaced from the microcomputer's RS232 level to a TTL signal), you will receive command data flawlessly all day long. As an experiment, hook a hobby motor to the same +5-volt logic supply from that bench that is supplying the power to the MC14469. The noise introduced by that motor will absolutely inhibit 90% of any further command transmissions from the microcomputer.

TTL signals also do not have the drive necessary to pump signals over long lengths of wire. In a rover system there may be 6 feet of wire between nodes. These same wires may travel past extremely (electromagnetically) noisy motors or other devices that would introduce spurious, unwanted negative noise glitches onto the data path. Some of these may be enough to simulate a valid address byte! A more valid, noise immune transmission path must be selected.

Two approaches to the design of noise immune communications links are presented in Fig. 1-23. The first (Fig. 1-23A) depicts using optical isolation as a means of decoupling any ground or signal originating noise. In most instances this approach will work. It is relatively inexpensive and does not significantly impact the producibility of the system. Optoisolators come in a number of varieties and are extremely low cost.

The second approach is more controversial. As shown in

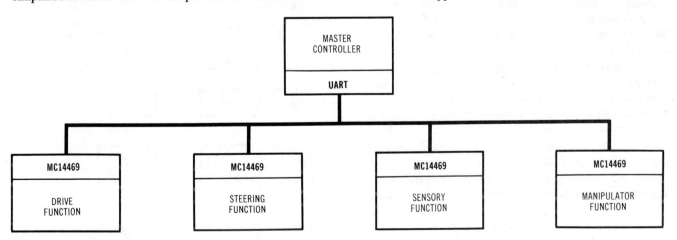

Fig. 1-21. Functional diagram of a rover system composed of several *function nodes*, each sharing a common communication link.

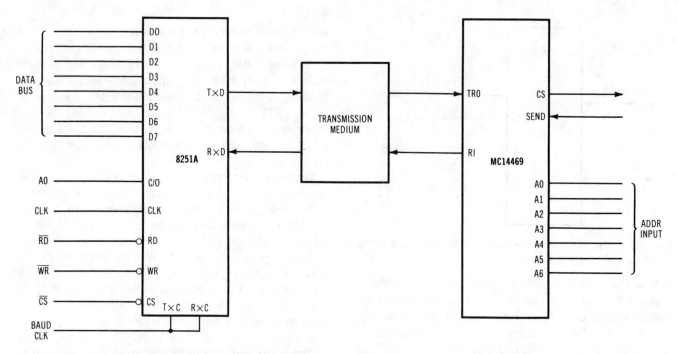

Fig. 1-22. Schematic diagram of the connection between master and MC14469.

Fig. 1-23B, the use of off-the-shelf fiber optic modules allows a completely noise-free communications path. Electromagnetic interference of the command cable will have no effect on the integrity of the data. The example using the isolators will still suffer effects of nearby motors. Using light as the transmission path eliminates all noise concerns. It does, however, bring up several other concerns.

The current cost of TTL-to-fiber interface modules puts them way above the amount necessary to simply optically isolate. The fibers used as the path must be cut to custom lengths and terminated properly, which is also extremely costly. Lastly, the party-line architecture will be dropped because of the unavailability of inexpensive T-type fiber interconnections. As noted in the figure, a star configuration would serve to be the best fit in an inexpensive fiber system.

I point out the use of fiber optics, amid so many contrary statements, because it is the connection media of the near future. When the cost and physical packaging dilemmas are solved, it will prove to be the best, by far, command link for electromechanical machines.

So far, I've shown you how the MC14469 will serve as a remote output node. What happens if you want to provide a node in the system that data will originate from? Surprisingly, the same part has the ability to transmit 2 bytes of information back to the master controller.

The complete transmitter portion of the part is illustrated in block diagram form in Fig. 1-24. This part cannot originate transmission to the master. If it could, it would cause a possible contention for the party-line data bus should another node be responding to the master at the same time. Instead, the transmitter is enabled only after a valid address byte is received from the master. In this way the master must poll the nodes asking for data.

Let's examine the sequence of events the part goes through for transmission. Upon the receipt of a valid address byte, the valid address latch internal status bit is automatically set. This will also enable the transmitter. The first byte transmitted out of the node will be its own address code that is hard wired onto pins on the package. The master can compare this

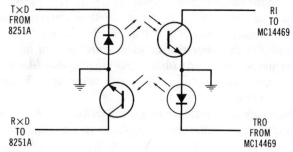

(A) Noise decoupling on the link, using optoisolators.

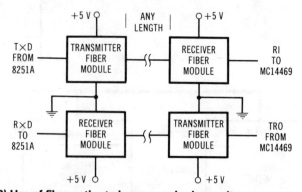

(B) Use of fiber optics to increase noise immunity.

Fig. 1-23. Designs of noise immune communication links.

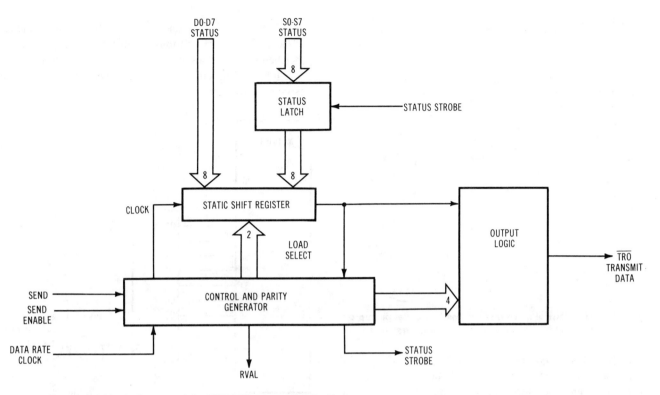

Fig. 1-24. Functional block diagram of the MC14469 transmitter section.

received byte against the address code it sent out to initiate the transmission to effect a crude error detection system. If the two match, allow the next byte in as valid data, or if they don't simply ignore it and try again.

In order to transmit these 2 bytes, the valid address latch must be set and the *send input* pin on the MC14469 must pulse high. This pulse would originate from an external circuit at the node. Obviously, the easiest implementation where, upon a valid address, the transmitter responds, would be to simply connect the valid address pulse output to the send input.

The second byte transmitted will echo the data presented externally to the send input pins S0 through S7 on the package. All 8 bits of this information are transmittable. What type of data would you transmit from a remote node such as this?

In a rover system, it might be necessary to measure some physical quantity. This type of operation usually requires that a conversion from an analog signal to a digital value be performed. The diagram shown in Fig. 1-25 depicts the MC14469 being connected to an 8-bit analog-to-digital converter. The valid address pulse, being connected directly to the start conversion input, will force the converter to accumulate a sample and convert the value to digital signals. When the conversion has finished, the converter will signal the MC14469 by pulsing its send input. This action will cause the 8-bit digital value, which is presented on the S-lines, to be transmitted as the second byte of the MC14469's message.

Another application for input processing is the example where the outputs are controlling a motor drive system (Fig. 1-26). You might want to add the capability of reading the value of the absolute position of the wheel as it is turning, which could be accomplished by using an optical pickup on the wheel. The detector might count the number of revolutions by shining a lamp on the wheel hub. A reflective spot would deflect the beam to the detector that produces a pulse per revolution.

These pulses would then be fed into a counter. The output of this counter could be connected to the S-lines. The valid address output signal pulse might then be directly connected

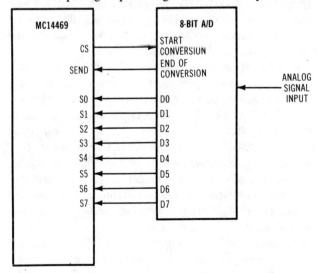

Fig. 1-25. Connection between the MC14469 and an 8-bit analog-to-digital converter.

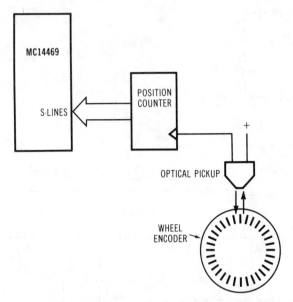

Fig. 1-26. Motor control node that utilizes an optical pickup on the wheel for keeping track of distance traveled.

to the send input (Fig. 1-27).

As you can see, the possibilities are limitless. There is one catch, however, when inputting data from devices such as converters. When the send enable latch is set after the receipt of a valid address, it will remain in that condition for only 8 bit-times. That means there is a finite amount of time that the converter must finish its conversion in. Some converters are relatively slow and unless the communication's rate that you chose is extremely slow, the conversion will not be completed in time. After the internal latch resets, the send input is disabled.

Combining both address and command functions with the transmit capabilities of the part will yield an extremely useful control node in a robot system. Fig. 1-28 shows this combination as an extension of the data converter design. Here an eight-channel A/D converter is selected, controlled, and read remotely through the MC14469. The channel select and start conversion lines are controlled via the control outputs, and the 8-bit data is read as previously discussed.

An entire robot system, utilizing this part is depicted in Fig. 1-29. The motion platform base unit is similar in design to the one shown earlier. The master control system or brain is shown here to be a Z-80 microprocessor; however, any microprocessor system can be used as long as it can transmit and receive a serial data stream in standard UART format.

The flowcharts of operation, as shown in Fig. 1-30 and Fig. 1-31, and the full timing diagram in Fig. 1-32 are provided to aid in the understanding of this part. It should be obvious that applying parts like these to systems can reduce interconnections significantly.

As I mentioned when I introduced distributed node processing, there are three vendors providing integrated circuits that are designed with this approach in mind. The Motorola, Inc., MC14469 was the first of the lot. Another remote part node-type circuit is manufactured by National Semiconductor Corp. This part, although similar to the previously described system, does exhibit a few unique qualities. Let's delve into the particulars of how and where to use it in the design of a rover system.

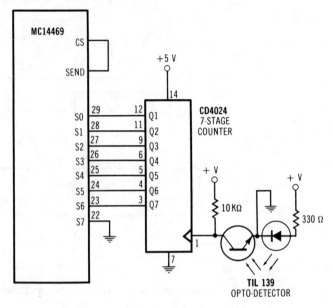

Fig. 1-27. Schematic of motor control node distance counter connection to the MC14469.

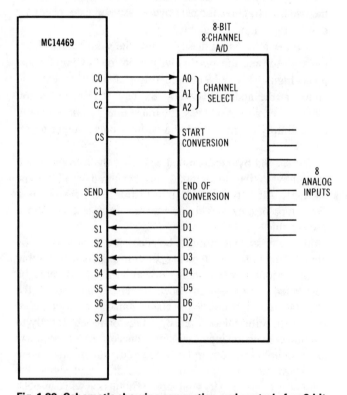

Fig. 1-28. Schematic showing connection and control of an 8-bit, eight-channel data converter.

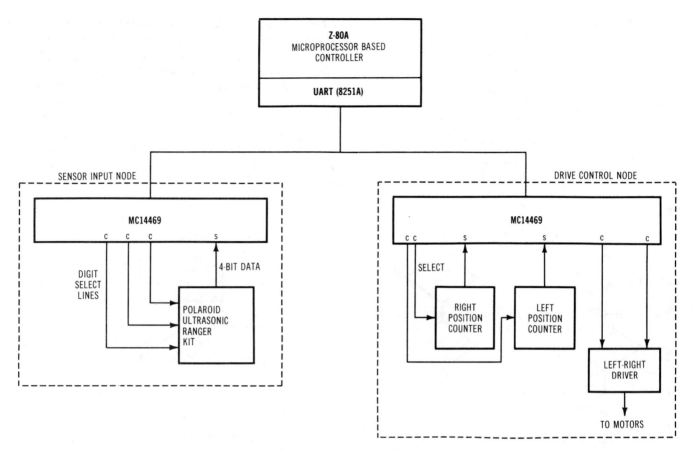

Fig. 1-29. Typical rover system implementation using the MC14469.

National Semiconductor Corp. MM54240 Asynchronous Receiver/Transmitter Remote Controller—This concept of remoting I/O by the use of a serial data stream has begun to catch on. In homes today there are little boxes plugged into the wall sockets that get addressed from a master control panel that is also plugged into a wall socket. It's this application that spawned the development of the MM54240.

Unlike the Motorola, Inc., part, this National Semiconductor Corp. system on a chip occupies only a 24-pin package. The differences don't stop there though, as the Motorola, Inc., unit could transmit and receive over a simple UART interface, the MM54240 requires one of its own to act as master. This is because the transmission link is pulse-width modulated. Fig. 1-33 illustrates the pin configuration and Fig. 1-34 shows the entire functional block diagram of the part.

Following the block diagram, we can examine its features. On the left of the figure we see the address inputs. As mentioned, this device may be configured as both a slave or a master. For the purpose of this introduction to the chip, let's stick to the slave mode. Later we will examine the master mode of operation.

Getting back to the diagram, the part will accept seven address inputs. Like the Motorola part, this too may answer to 1 of 128 addresses. In slave mode these are hard wired to the address code you would like it to respond to. The master sends out a message that consists of the following:

7 Address bits
1 Command bit (read = 0, write = 1)
8 Data bits
1 Parity bit (even parity)
1 Dummy bit

There are no valid address pulses or the like, however, the CS pin, which in the slave mode acts like an output, can signal a read or write operation. The interface between the MM54240 and external circuitry is one 8-bit bidirectional port. The CS line would be used to enable buffers external to the part. These eight lines may be configured, by the user, into four possible combinations of functions.

Fig. 1-35 lists the connection possibilities for the two output configurator input pins. The modes listed there allow eight outputs, a combination of four inputs and four outputs or a specialized mode where the logic zero outputs are low-impedance output ports yet the logic one outputs have a weak pull up to +5 V to allow wired-OR conditions with external circuits.

The major difference or disadvantage this part has in comparison to the Motorola unit is the fact that it can only capture 8 bits on one message. We have seen where the other part can both transmit a 7-bit word and latch it up internally while receiving an 8-bit reply. The National Semiconductor Corp.'s part requires external circuitry to accomplish this.

Another apparent disadvantage is the necessity of using

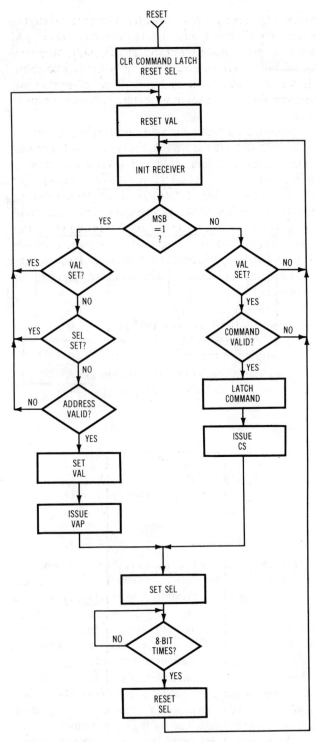

Fig. 1-30. Receiver operation flow diagram.

Fig. 1-36 depicts a typical master-mode hookup between the device and a standard microprocessor bus. A nice feature is that the address lines can be connected directly to the microprocessor's address bus, and each of the 128 remote nodes may be addressed like memory locations. In a Z-80 processor application, the I/O space instructions would work well there. Fig. 1-37 illustrates an I/O mapped application.

Status information may be read from the part; however, its implementation is less than standard for microprocessors. An input signal, separate from the others (S-pin 16), when taken to a logic low level, will cause a 3-bit code to be output the same device as master. In the case of a robot controller, this would require the main microprocessor to interface with the MM54240. This type of interfacing is called the *master mode* and is determined for the chip by allowing the mode input (pin 11) to float. In this application the CS input becomes a standard *chip select* and the two output configuration inputs C1 and C2 become *read* and *write* control inputs from the master microprocessor.

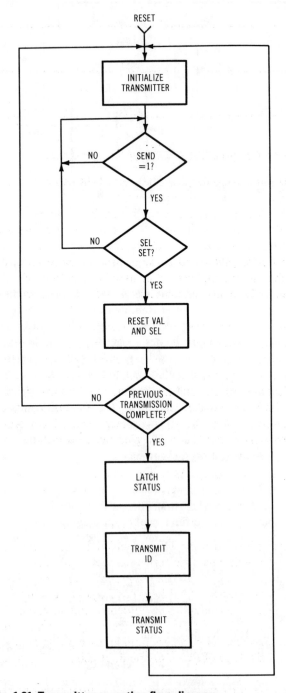

Fig. 1-31. Transmitter operation flow diagram.

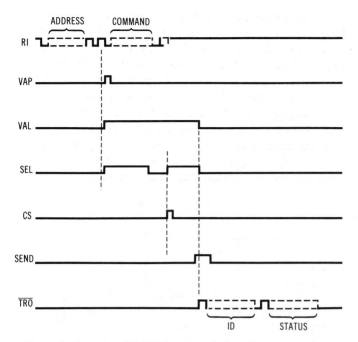

Fig. 1-32. Complete MC14469 system timing diagram.

on the three low-order data lines. Fig. 1-38 lists each code and the associated status indication.

What does the serial pulse-width modulated output consist of? Well, that takes us into a discussion of the internal oscillator of the part. We did not cover the on-chip clock circuits of the MC14469 in the last section since showing both that and the implementation used in the National Semiconductor Corp. part simultaneously would be more instructive.

Fig. 1-39 illustrates both clock oscillator configurations and the required connections. As you can see, the Motorola Inc., part requires a ceramic resonator tuned to a specific frequency. Shown is an implementation for 4800-baud communications. An external clock, generated by some circuit, may also be fed into the OSC1 input. The National Semiconductor Corp. allows the use of an inexpensive RC combination to achieve oscillation. To pick the combination, you simply use the following formula:

$$V = +5\text{ V} + 5\%$$
$$F \sim 400\text{ kHz}$$
$$F = K/RC$$
$$K = 0.8 - 1.4 \text{ (depends on part)}$$

Getting back to pulse-width modulation, the diagram in Fig. 1-40 depicts the waveforms for both a data zero and a data one condition. One bit is equivalent to 96 clocks of the oscillator frequency. If you keep this frequency at 400 kHz, then one bit is equal to 240 microseconds and the entire 18-bit message would take 4.32 milliseconds to transmit. The figure shows the width percentages for each bit state. In all cases, the voltage output of the part is at standard TTL-level signals.

Because of the fact that these take a considerable amount of time to transmit to a slave, a master microprocessor trying to read data will have to first initiate the read command, and then read status to see when the data is ready, then read again. This can get somewhat hairy. It is best to accomplish this period of waiting in hardware. If the microprocessor used incorporates a ready or wait input then the problem is solved.

A hardware interface utilizing the wait feature is depicted in Fig. 1-41 and both the write operation and read operation timing diagrams of Figs. 1-42 and 1-43 should serve to illustrate and clear up any implementation questions. For both the Motorola, Inc., MC14469 and the National Semiconductor Corp. MM54240, I suggest you contact their local representatives or their factories directly before doing any serious design tasks with these devices. They can provide you with detailed specifications and, in some cases, applications notes on using the part. I offer their addresses for your convenience.

Motorola Semiconductor Products, Inc.
3501 Ed Bluestein Blvd.
Austin, TX 73721
Part number MC14469 (CMOS Data Book)
Ap note AN-806 *Operation of the MC14469*

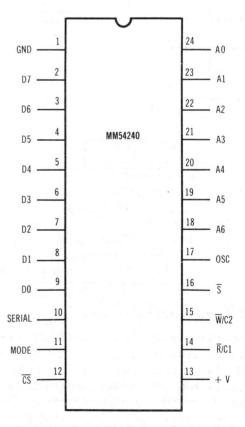

Fig. 1-33. Pin designations of the National Semiconductor Corp. MM54240.

29

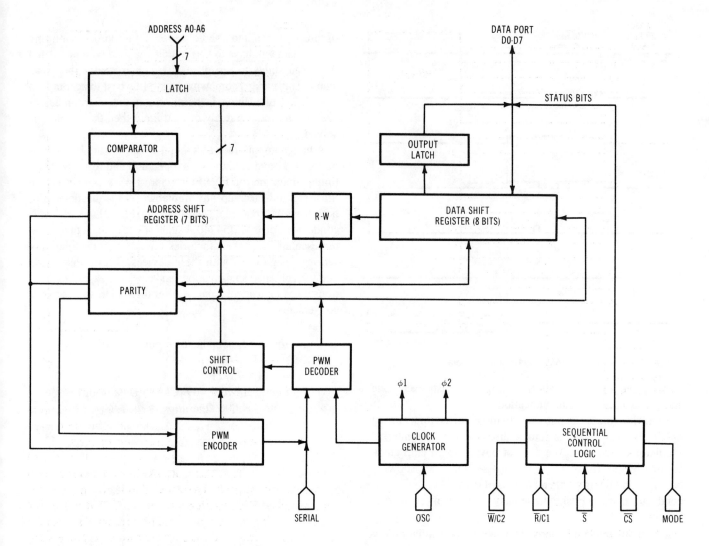

Fig. 1-34. Functional bock diagram of the MM54240.

National Semiconductor Corp.
2900 Semiconductor Drive
Santa Clara, CA 95051
Part number MM54240 (MOS/LSI Data Book)

Let's finish off these two devices before starting into the third and final intelligent node component because, when you examine this part, you will finally see the power of distributed processing. The two previous parts, although cutting down on system interconnections, did little to relieve the master processor from many mundane duties. This part, manufactured by Mostek Corp., will prove to act not only as a slave port, but can make decisions on its own, therefore, relieving the master of some number of duties. Let's begin our exploration.

Mostek SCU20 Serial Control Unit—This device is a custom-programmed single-chip microprocessor. The program that has been masked into the part allows it to function as an intelligent control node in a serial party-line drop system. Fig. 1-44 is a functional block diagram of the internal structure of the part.

Its features include two 8-bit bidirectional data ports. A third port may be used as either another bidirectional data port or as the address code input port. There are six timer/event counters that may be started, stopped, loaded, or read via the serial link. A set of instructions exists that allows the master controller to interrogate various functions or command the part to perform complex actions.

Let's investigate each of the part's functions one by one.

C1	C2	DESCRIPTION
1	1	All eight pins are high-impedance input ports
0	1	All eight pins are standard low-impedance output ports
0	0	D1–D4 are standard low-impedance output ports
		D5–D8 are high-impedance output ports
1	0	Logic 0 outputs are low-impedance output ports
		Logic 1 outputs are weak pull-ups to +5V

Fig. 1-35. Output possibilities and their corresponding configuration input code connections.

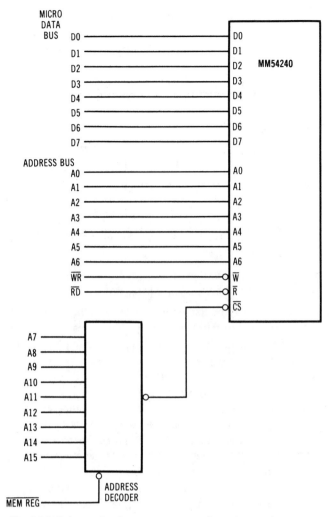

Fig. 1-36. Schematic of memory mapped master-mode connection.

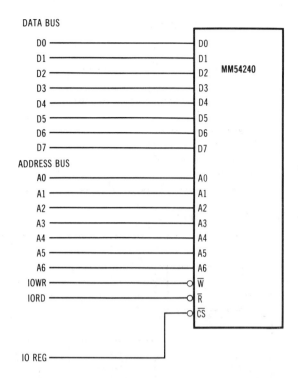

Fig. 1-37. Schematic of I/O mapped master-mode connection.

port bit is to be used as an output. Conversely, a logic zero will program that port bit to act as an input. Therefore, each 8-bit port can be configured in a multitude of ways.

Ports 4 and 5 are general-purpose ports. Port 4 also includes a strobe output that will pulse low whenever the port is loaded with data from the master. This strobe may be used by external circuitry to trigger some event. Port 5 does not contain this feature. Port 0, as mentioned, may act like the other two; however, the designer must make a choice when using this port.

Addresses of slave SCU20s are set in one of two ways. Port 0 may be used to read in a hard-wired address code, or the code may be automatically loaded upon power-up through a special serial data port. Both methods allow a code that depicts up to 255 slave devices, that is, almost twice as many as the previous two parts. Selection of which method to use is done by the use of the SERADIN input (pin 37).

In that way a better understanding of its application is possible. The SCU20 receives and transmits over separate serial data lines. The signals used are TTL levels, and the format is compatible with standard UARTs. The rate at which transmission is received and transmitted is selectable by connecting certain levels to the baud rate select inputs of pins 24 and 25 (see Fig. 1-45 for pin assignments). The chart in Fig. 1-46 describes the specific codes and the rates they select.

Because of the separate lines, the interconnection between master and all slaves is called half-duplex. Fig. 1-47 illustrates a typical hookup of the communications link. Messages sent over this link are much more complex than the previous two examples. Because of this, we will limit the discussion of this part to hardware functions in this chapter. The next chapter will discuss, fully, the command structure and message functions of this part.

Looking back at the block diagram, let's examine the structure of the 8-bit ports. Each port has an associated data direction register. These separate registers serve to configure each bit within the port. A logic one loaded into a bit position of the data direction register will indicate the corresponding

| STATUS REGISTER | | | DESCRIPTION |
D2	D1	D0	
0	0	0	Not used
0	0	1	In process of transmitting to slave during write mode
0	1	0	Valid data received from slave
0	1	1	Not used
1	0	0	Awaiting data from slave during read mode
1	0	1	In process of transmitting to slave during read mode
1	1	0	Invalid data received from slave
1	1	1	Initialization/idle condition

Fig. 1-38. Status codes output to the bus when S (pin 16) is brought low.

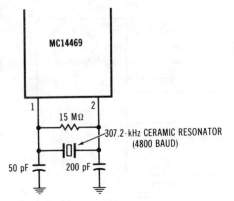

(A) Motorola, Inc., MC14469.

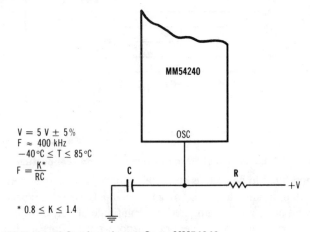

(B) National Semiconductor Corp. MM54240.

Fig. 1-39. Clock oscillator configurations.

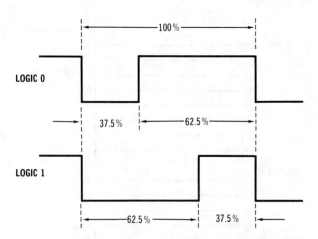

Fig. 1-40 Output wave forms of the National Semiconductor Corp. MM54240 part.

If SERADIN is tied to ground (Fig. 1-48), the code hard wired to port 0 will be loaded as the address of the slave. This method limits your use of ports to two, since port 0 can no longer be used as a general-purpose port. If the other method is chosen, it is necessary to serially load the address into the SERADIN pin. The SCU20 provides the necessary clocks and strobes; all you have to do is connect an external parallel-to-serial shift register component as shown in Fig. 1-49. As you can see, the addition of a single 16-pin component will free up a tremendous amount of functionality.

Each port may be loaded with data from the master, and data connected to those ports may also be read. There are times that only certain bits, within a port, are important, and the master will usually read in all eight then mask off the unimportant bits by ANDing them with zeros. This operation can be time consuming so the first way the SCU20 can off-load processor intervention is to include this function in its own hardware. Each port has an associated mask register that may also be loaded and read by the master. There are certain read commands that automatically use these registers.

To stop at this functionality would probably be fine for many applications. It certainly outperforms the previous two parts. Moving up the block diagram, let's move left toward the timer/counters. The description of the MC14469 mentioned an application using a counter in motion control. Can you guess what this part has the ability to do?

As you can see, there are six identical event counters. Surprisingly, they are 15 bits wide. Each is an up-counter that starts at a value of zero. They may be incremented by one of three sources. An internal timer oscillator can increment

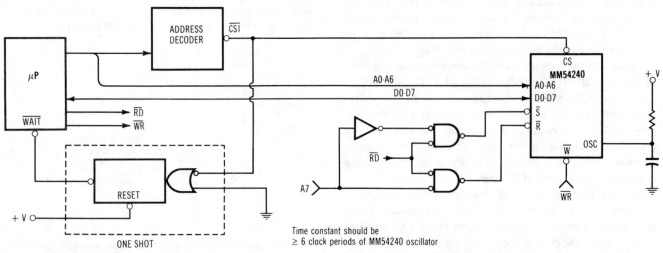

Fig. 1-41. Hardware interface utilizing wait logic.

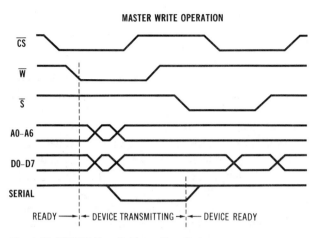

Fig. 1-42 MM54240 write operation timing.

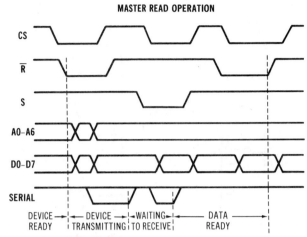

Fig. 1-43 MM54240 read operation timing.

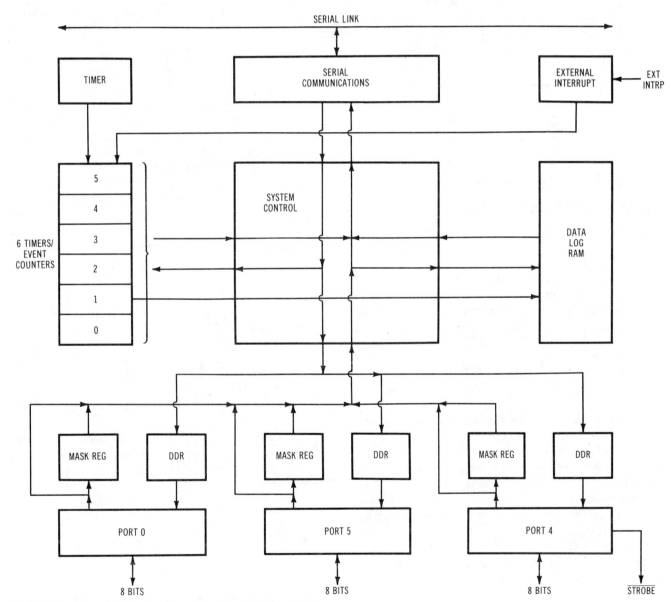

Fig. 1-44. Functional block diagram of the Mostek Corp. SCU20.

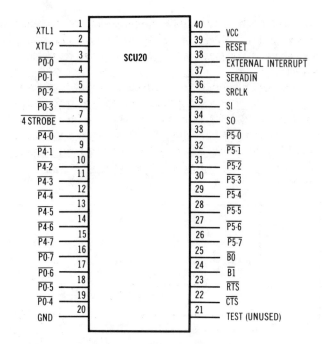

Fig. 1-45. Pin designations of the SCU20.

B0	B1	BAUD RATE
0	0	300 BAUD
0	1	1200 BAUD
1	0	2400 BAUD
1	1	9600 BAUD

Fig. 1-46. Communication data rate select input code chart.

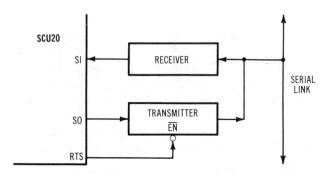

Fig. 1-47. Connection diagram showing two-way half-duplex communication link between parts.

the counter once per a 10-millisecond period. This is useful for timing events. Another way of incrementing is to have the master control issue a *step event counter command*. The third way to trigger the counter is through the *external interrupt input* (pin 38).

The master has the ability to clear, start, stop, or read each counter. A value may not be preset into these from the master; however, by selectively incrementing a counter, using the step counter instruction, a value may be loaded. This method may seem time consuming, but it's the only method available. There is one command that does allow a limit value to be loaded into the event counter logic; however, this only applies to counter 1, and I'll go over that use separately.

Using a counter for determining position is easy when the external input mode is opted for. You could connect the SCU20 up as shown in Fig. 1-50. This application controls two wheels and detects the distance traveled by the rover. Each detection of the reflective dot on the wheel will increment the internal counter. The master controller needs only to clear the counter, start it in the external mode, select the motor control outputs, and periodically read the counter value looking for a specific value that indicates the distance traveled so far. At the point when you have arrived at a preset distance, the controller would issue an I/O port command that would stop the motors. There are several hundred other uses for counters like the ones provided in the SCU20. In the

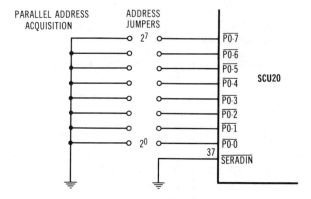

Fig. 1-48. Use of Port 0 to select device address.

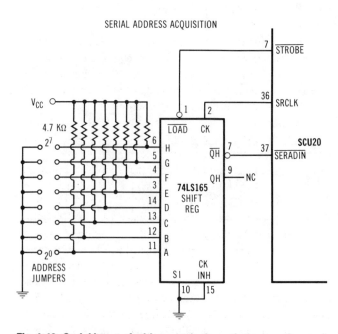

Fig. 1-49. Serial input of address code through the use of a parallel-to-serial shift register.

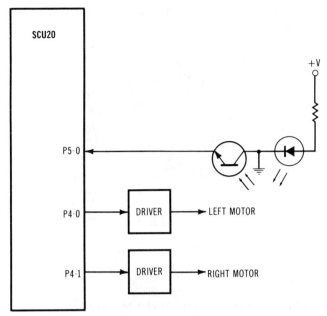

Fig. 1-50. Schematic diagram showing SCU20 being used for motor control.

next chapter, the commands associated with the counter section will be explained fully.

Data Logging

As I mentioned when explaining the counter facilities, counter 1 has an ability that no other counter has. The block diagram shows an area within the chip that is called *data log RAM*. The most intelligent duty that the SCU20 has been programmed to do is to automatically *log* data from one of its three ports. Up to 63 data samples may be automatically taken and stored in internal memory upon the issuance of a command.

Data logging begins when counter 1 reaches a preset value. That value is part of the argument of the *start data log command*. Each time that value is reached another sample is taken. You may specify any port (0, 4, or 5). The strobe pin will go active after the logging has started and will continue pulsing after each sample. This is useful as a handshaking signal to whatever external hardware you are sampling from. Each sample is stored in its own location in memory.

The master has the ability to read the count of samples and stop logging at any time. After 63 samples have been taken, logging automatically stops. It should be noted that while in the data log mode, the master may use any other SCU facility (ports, counters) without affecting the operation of the log.

Analog converters may be connected and several samples over time can be stored. Remember that the counter used can be incremented to that *trigger value* by one of three methods. Fig. 1-51 shows a possible application using a digital camera that outputs in bytes. Similar devices may be connected.

As you can see, the SCU20 becomes our true distributed processor. Of course, it is very possible to design your own application-oriented processor using a single-chip microprocessor such as the 8048. Some of the newer parts (Motorola, Inc., 68701; Intel 8051) include UARTs on chip that would make them ideal additions to an SCU-based system.

Rover controllers may be constructed in a variety of ways, as you have just seen. A systems approach to their design must include both the function they are to perform and the hardware required to implement that function. We now move away from the control block on the master systems diagram and over to a side input called *power systems*.

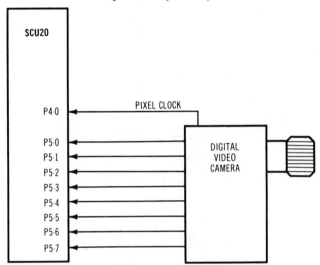

Fig. 1-51. Diagram showing the use of the data log facility. Pixel clock increments internal SCU20 counter through port 4.

Power Systems

The electricity required to power a rover system depends on the application. If the vehicle is very small, is lightweight, moves short distances, and doesn't have a lot of "bells and whistles," then primary batteries may suffice. However, in applications where everyday use may drain the power system to nothing, a type of rechargeable secondary battery would be required. Let's briefly look into the differences, terminology, and applications of various power sources for rovers.

If you haven't had this material since high school, it should serve to be a pleasant refresher. All battery or portable power systems, excluding solar, rely on the *voltaic cell*. The principles by which it operates are relatively simple. Refer to Fig. 1-52 while you read the description of its operation.

The cell consists of a container that is filled with a substance called *electrolyte*. This substance can be in liquid or wet paste form. It is made up of either a salt, alkaline, or acid solution. Inserted into this chemical are two metal electrodes. One is made of carbon and represents the positive pole of the battery. The other consists of zinc. In standard flashlight batteries, the zinc electrode is actually the container in which the electrolyte paste is sealed. The positive carbon electrode is inserted into the center of the container

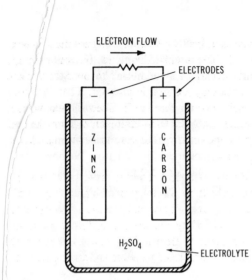

Fig. 1-52. Basic voltaic cell.

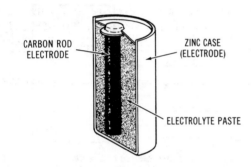

Fig. 1-53. Construction details of a standard flashlight cell.

and insulated from the zinc can. Fig. 1-53 depicts the construction of a standard flashlight cell.

These types of power sources are called *primary cells*. The chemical reactions that produce the electricity eventually erode both electrodes; however, the negative one sustains the most damage, and voltage output drops. Primary cells are throw-away items. In cases where a rover draws so little power that the lifetime of the cell will approximately equal the shelf life of the battery (about four years for alkaline), this type of system would be ideal. They are relatively lightweight and come in a variety of voltage ranges, capacities and physical configurations. The standard primary voltage values are 1.5, 6, 9, and 12 volts. Some of the newer mercury batteries, used for smoke detectors, etc., allow some nonstandard voltages like 4.5, 11.2, and 12.5. There are a number of specialized batteries that can be purchased that are several standard cells connected in parallel and series to effect the desired voltage.

Secondary batteries have electrodes that, although damaged somewhat by the chemical reactions, can be recharged. This recharging is done by forcing current flow through the battery in the opposite direction from the normal current flow. Nickel-cadmium batteries are currently being used in shavers, certain flashlights, some toys, and most calculators.

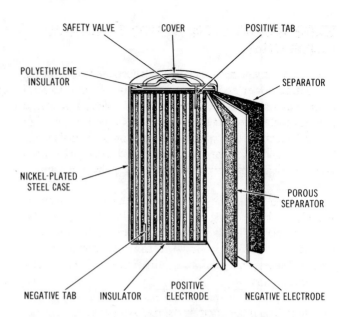

Fig. 1-54. Construction details of a standard nicad cell.

The construction of nickel-cadmium (nicad) cells is substantially different from lead-acid batteries. The drawing shown in Fig. 1-54 illustrates the mechanical details. Notice that there are several layers of material. The positive electrode material is a nickel hydroxide and the negative is made of cadmium hydroxide (makes sense doesn't it?). The separator is an absorbent nylon material that soaks up the electrolyte.

Recharging a nicad cell is relatively simple. The circuit of Fig. 1-55 shows all that is necessary. This circuit will charge cells at a 45-milliampere rate. After an overnight charge, your rover should be as good as new. It should be noted here, that you should investigate power sources more fully than I can describe in these pages. There are several books devoted to the subject of batteries, and most basic electronics coursebooks have a chapter on it. For a good lesson in power sources, as they relate to robotics in particular, you should check out obtaining the robotics course from Heathkit®*. The following are the particulars:

>Heath® Co.
>Benton Harbor, MI 49085
>Course #EE-1800
>Robotics and Industrial Electronics

Having gone over all the blocks necessary to create a functioning rover system, it is time to concentrate on a few real-life applications. The following is an account of the design experiments of two rover systems. Each is different and documented as fully as possible. These examples are shown here to stimulate your thinking and perhaps further illustrate the design of a rover system.

---

*Heath and Heathkit are registered trademarks of Heath Company.

EXPERIMENT 1: M1 REMOTE, MOBILE PLATFORM

An easy entry point to rover system's design is to use existing computing power and add a remote drive system. Control algorithms for guidance, navigation, and collision avoidance may be developed without the need for expensive hardware. M1 is a first attempt at this type of a system. Composed mainly of a standard toy-store variety radio controlled truck, M1, when properly interfaced, can perform with the agility of a complex robot system.

Fig. 1-56 is a photograph of the vehicle. It is approximately 14 inches long by 7 inches wide. This particular truck chassis is manufactured in Japan for the Shinsei Corp. of Cenitos, California. Its trade name is Mountain Man Chevy Blazer. This four-wheel drive vehicle comes with a hand-held transmitter unit that includes two joystick controls. One joystick controls the speed and forward or reverse direction. The other controls the right-to-left steering. Both controls will operate the steering and speed functions proportionately to the position of the stick.

On the transmitter, both joystick controls are marked with several steps. The sticks normally sit in the center of the range. We will call this the "null zone" because the vehicle is set at rest when they are in this position. There are four markings in each direction from null. These indicate offset values from center. In order to interface the transmitter unit to an existing computer, it will be necessary to emulate the action of the joystick.

Let's examine the electrical representation of what each joystick control presents to the transmitter circuit. Fig. 1-57 is the schematic diagram. Notice that there are two variable resistors per control. They are each a different value of resistance. It is possible to electrically emulate the action of each of these pots by using fixed resistors and analog gates or relays.

Examine the schematic in Fig. 1-58. Here is shown an equivalent circuit to the null point of a 10-kΩ pot. Notice that each relay must be on to complete the circuit. Adding resistors in parallel to these (as shown in Fig. 1-59) and then selecting different combinations of relays will effectively

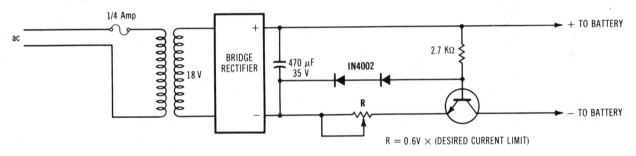

Fig. 1-55. Simple "no frills" nicad battery charger.

Fig. 1-56. Photograph of M1, a remote-activated mobile platform.

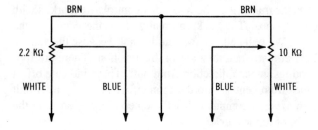

Fig. 1-57. Schematic diagram of the joystick section of the transmitter.

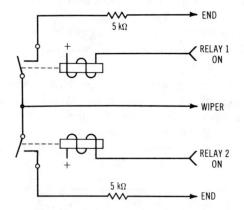

Fig. 1-58. Representation, using relays and fixed resistors, of the mid-point of a 10 kΩ variable resistor.

simulate the action of the variable resistance; however, it will vary in steps as opposed to a linear fashion. In practice, having all those relays consumes much power and is also somewhat bulky.

In the last volume, I explained the operation of the analog gate. Briefly, this part consists of a pair of transistors that, together, can pass a varying signal or voltage almost as reliably as a relay. However, this switch does have a significant internal resistance of its own. The reason to mention it here is that these components come in configurations that allow for the simulation of circuits, such as the one in Fig. 1-60, with a minimum of packages and power dissipation.

The CD4052 is a dual four-channel analog gate. With this part, the circuit of that figure can be reduced to the one shown in Fig. 1-61. Here, resistance values are chosen according to the 2-bit digital signal applied to the analog selector. In this experimental vehicle, however, only three of the positions are to be utilized. With the vehicle at rest, the null position must be selected. In order to determine the values of resistance for that position measure the positions shown in Fig. 1-62. After getting these, the resistance at one point on either side of the null is chosen.

I have found, that due to varying resistances within the CD4052, it is better to affix variable pots to the inputs instead of fixed resistors. In the end, the schematic looks like that of Fig. 1-63. After a little fine tuning, you should find that

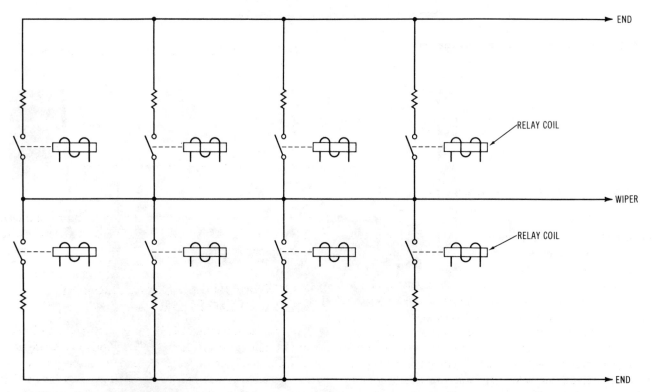

Fig. 1-59. Circuit showing the use of many relays and resistors to selectively vary (in steps) the pot representation. Circuit requires eight relays and associated control inputs.

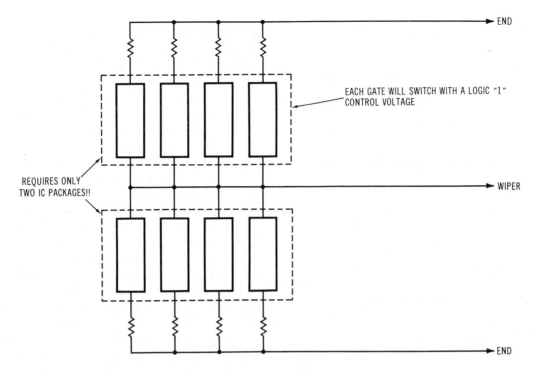

Fig. 1-60. Schematic showing the use of analog gates to replace the relays used in Fig. 1-59.

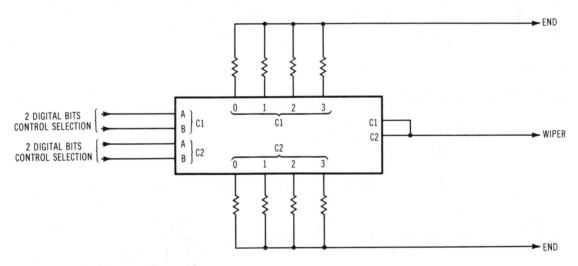

Fig. 1-61. Use of the CD4052 in replacing a variable resistor.

connecting the two bits to a computer output port will produce a similar joystick action. An identical circuit is used for the other control joystick. With one 4-bit word, you can emulate the human action of the joystick.

Moving back to the vehicle, there are no changes necessary to the circuitry. It will receive commands for steering and speed just like it did with a human operator. There is one requirement that I added. Sometimes it is nice to know when the vehicle has hit an obstacle. A bumper, which activates a switch, is simple enough. However, getting this information back to the remote computer must be accomplished via radio.

I salvaged a simple on-off type radio car transmitter and wired the on switch to the bumper. The receiver incorporated a relay that, when pulled up to +5 volts, results in a nice TTL pulse. The schematic for this feedback system is also included here in Fig. 1-64.

EXPERIMENT 2: M2 INTELLIGENT MOBILE PLATFORM

As the name implies, M2 includes an on-board computer. In fact, this vehicle more closely resembles the descriptions of the rover systems presented earlier. Although only an experimental model, M2 has proven to possess interesting unique qualities. Let's examine the architecture of the vehicle.

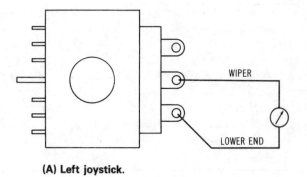

(A) Left joystick.

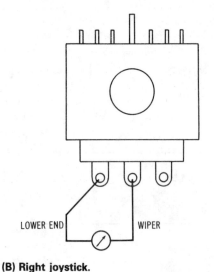

(B) Right joystick.

Fig. 1-62. Figure showing the positions of the joystick and the terminals to be measured at these positions.

Fig. 1-65 is a block diagram of the various functions. Notice that this diagram is somewhat detailed as it gives the particulars of the functions. An RCA 1800 family microprocessor was chosen as the master controller due mainly to its CMOS low-power structure and its ability to handle I/O in an extremely efficient manner. An explanation of the inner workings of the 1800 family, as it applies to M2, is given later in this chapter.

Looking back at the block diagram, we see that M2 is composed of the 1800 microprocessor as a controller; a separate drive control board; another separate sensor input, which is called the ultrasonic ranging system; and a separate function called the power system. This is, as you can see, a very simple rover system implementation. Other things could be added to the general bus structure, such as various output devices, manipulators, and other robotic items. M2 is used as a test bed for different input and output systems as well as for the use of developing software that will allow it to maneuver a room without any human intervention. The controller portion of M2 is based around a single-board

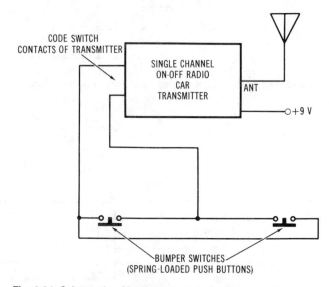

Fig. 1-64. Schematic of bumper remote feedback system.

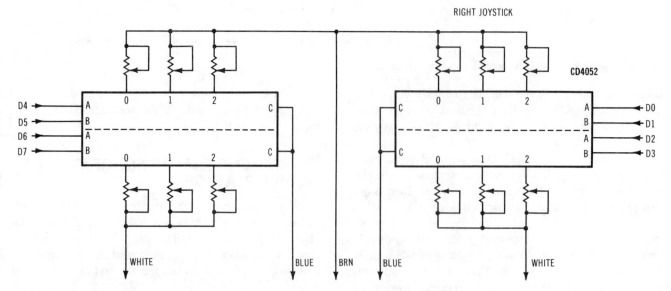

Fig. 1-63. Schematic of the remote transmitter interface.

computer that became available in the mid-to-late 1970s. The board was manufactured by RCA and was called the COSMAC VIP Series Home Computer. On this board was located 4K of CMOS memory, a general-purpose 8-bit input/output port, a video output circuit that allowed a standard television type monitor to be connected that would display memory locations, a 16-key hex keypad, and a preprogrammed ROM that contained a very basic debugging utility. M2 utilizes all of this functionality. Of course, you could build your own single-board controller. The debugger and how it works within the M2 system will be expanded upon later. Right now, let's get into the 1800 family of microprocessors to explain their workings within M2.

As shown in Fig. 1-66, the functional block diagram of the 1805 microprocessor shows that it is a very simple architecture. There are many advantages to an architecture this simple. It is based mostly on a register array composed of 16 general-purpose 16-bit registers. Each register can be used to hold data or address locations. Each of these registers is designated by a 4-bit binary code known as a *hex notation*; therefore, with 16 registers, each register is designated 0 through register F. You will notice from the block diagram that it is an 8-bit microprocessor. There is an 8-bit arithmetic logic unit and a date register, which would be similar to the accumulator in most microprocessors. Memory addresses, at the top of the figure, come from a multiplexer that allows 8 bits of address to be output at one time. This type of structure is known as a multiplexed bus. This cuts down on the pin count and allows for many I/O function pins that take the place of the extra eight address lines. The 1805 has the capability of addressing up to 65,536 bits of memory external to the part. This would result in 16 address lines. By multiplexing these lines, only eight pins are necessary.

There are two timing signals. One called TPA, which signals that the first eight address bits or the high-order address byte is on the address bus. TPB signals that the low-order byte of address is on the bus. That's what the multiplexer is used for. If you'll notice in the diagram, there is an increment/decrement circuit. When we go through some of the instructions, you will see that it is possible to automatically increment and decrement memory locations. There are a number of small 4-bit registers within the part. The N register, P register, X and I registers are used for control functions. The N register is part of a very complex I/O structure that will select one of eight banks of I/O. There are three N lines on the part and the output of this N register will show up directly on these pins.

Now is a good time to show the pinout of the 1805. Fig. 1-67 is a representation of the pin designations. The P register will dictate to the system, which of the 16 general-purpose registers is to act as the program counter. For those of you familiar with microprocessor architecture, you will immediately note that this becomes a very powerful system. Within a 1-byte instruction, changing the P register can change the operation of the microprocessor.

In the event of an interrupt, you can immediately switch to a different location in memory probably faster than any microprocessor of this type available. The I register is an extension of an instruction register, where a portion of the instruction is stored. The I portion happens to be the high-order digit of the instruction format and, normally, indicates instruction code types.

The N register is also used, in this case, to store the low-order 4 bits of an instruction byte. As I said in the case of the I/O instructions, this will select which bank. In some other instructions, this may be some variable that indicates a subset

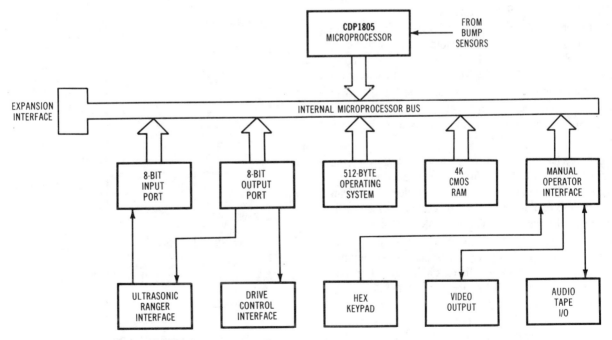

Fig. 1-65. Functional block diagram of M2.

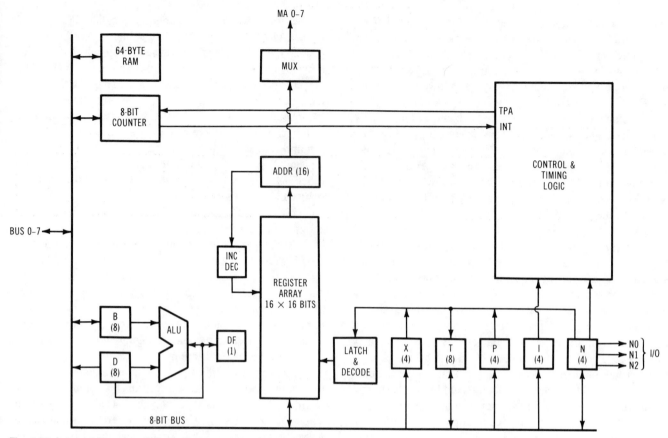

Fig. 1-66. Internal functional block diagram of the CDP1805 microprocessor.

of an instruction code. The X register is usually used to point to a destination or a source within an instruction, where the next byte may be taken from a certain register. X will point to that register. There is a temporary 8-bit register called the T register, and as mentioned earlier, the D register acts just like an accumulator in most microprocessors.

Up at the center top of the diagram (Fig. 1-67), you will see a small block called Q. This is an interesting point. The Q is a latchable output bit. It may be used for any purpose. There are many instructions that support Q, and you will see the Q output being used to strobe data out of the rover system. Last but not least, there is an 8-bit general-purpose timer/counter, which you will also see used in the rover system.

From here, let's go on to the entire architecture of the single-board 1800 controller. Fig. 1-68 is a functional block diagram of the VIP board, used as the controller. As you can see from the figure, the 1805 microprocessor is being used to coordinate all the functions. There are 4K of CMOS memory, a built-in hex keyboard, the graphic video display interface, an audio cassette interface, clock generation circuits, and a beeper. Notice, also, that there is a 512-ROM operating system, which is the debugger and a general-purpose parallel I/O port and a separate expansion bus connector.

The board interfaces to the rest of the system, as shown in Fig. 1-69. In this figure, you will see that, through the I/O port connector, the motor control board and the ultrasonic ranger board are connected. This is done by utilizing portions of the output byte to select and control each board. From here, let's go on to the motor control board.

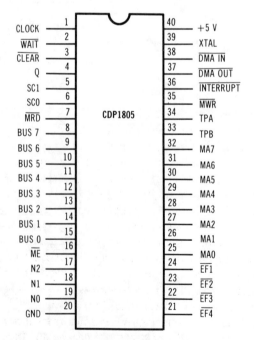

Fig. 1-67. Pin designations of the CDP1805.

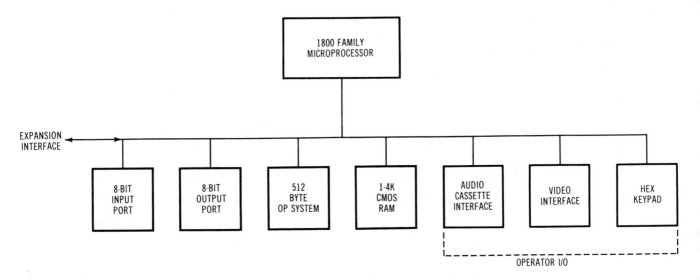

Fig. 1-68. Functional block diagram of VIP controller board.

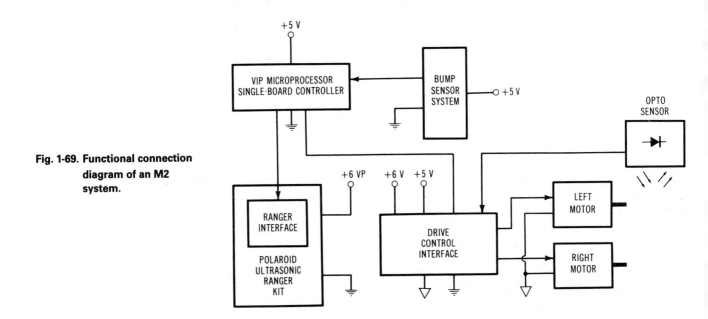

Fig. 1-69. Functional connection diagram of an M2 system.

The motor control board is relatively simple. The block diagram in Fig. 1-70 shows that there are three relays on the board. One to control the power to the motors and two others, one to control the direction of each of the two motors in the base. There are drivers on board that are used to handle the current of these relays, and there is a single latch that receives commands from the VIP output port. As you can see, there are 3 bits associated with the motor control portion. There is 1 bit for the direction of each of the motors and a separate bit for enabling the entire drive subsystem. Data is latched into the output port by the use of the Q output of the microprocessor. Typically, the software will select direction changes, then enable the motors. In that way there are no changes within the operation of a motor from forward to reverse. This way, the software will disable the drive por-

tion, change directions, and then enable it, which also effects a short pause in the motor. A full schematic diagram of the motor controller board is shown in Fig. 1-71.

Also utilizing this output port is the ultrasonic distance ranger. With this interface, M2 is able to locate objects in its path through the use of high-frequency ultrasonic sound waves. I went over the operation of ultrasonic systems, including the Polaroid ultrasonic ranger kit that is utilized in the M2 design in the first volume of this series. Suffice to say here, that the only difference between the M2 utilization of the kit and that shown earlier is the interface. The interface between M2 and the kit is shown, in schematic form, in Fig. 1-72. Basically, what this interface performs is that it allows the controller board which of 3 bytes of distance it wants to read. As you can see from the schematic, the output of the

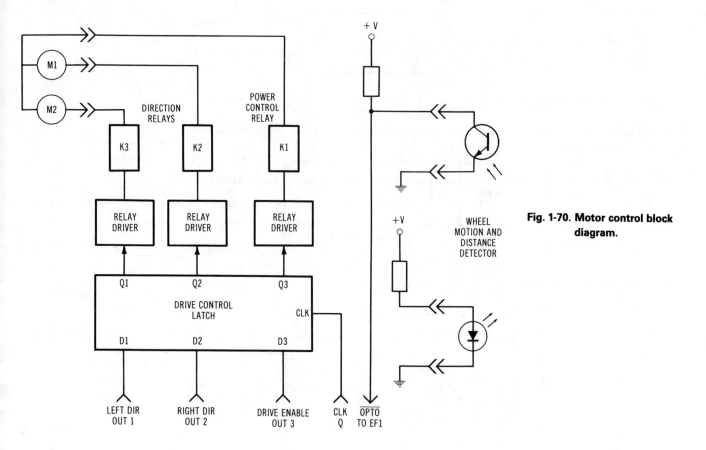

Fig. 1-70. Motor control block diagram.

ranger, where we pick it up, is a 4-bit binary code. These codes are going by three digits at a time depending on the outputs of the three control outputs shown in the diagram. With the ranger kit you can detect from 0.9 feet to 35 feet in distance. This would be shown as a 009 to 350 in the three digit positions. Through the use of 2 bits we can select which one of those three digits we would like to read and what the 4-bit code will be, automatically, latched into the input port section of the VIP control board.

As I said in the beginning, I was going to show the basic circuitry of both M1 and M2 just to stimulate your imagination. You will gain a much better understanding of the operation of both of these experimental vehicles throughout the pages of the following chapter. It is there that the operational and software aspects of designing rover systems are covered.

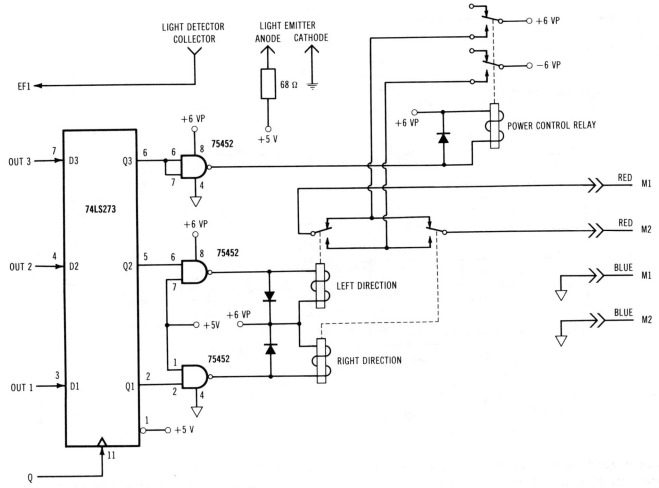

Fig. 1-71. Schematic diagram of the motor controller.

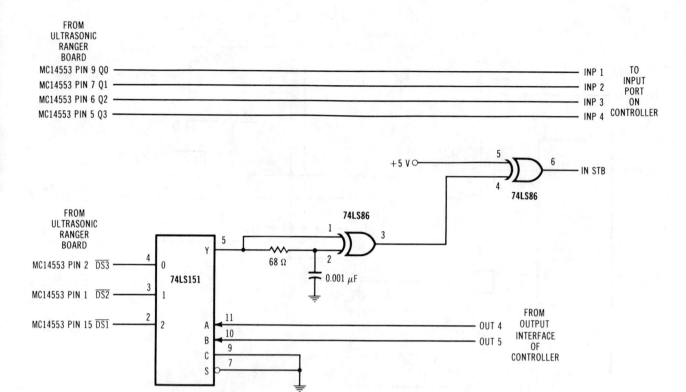

Fig. 1-72. Schematic diagram of the ultrasonic ranger interface.

Chapter 2

Software for Rover Systems

With the advent of low-cost computing power, traditional hardware-intensive systems have been redesigned to take advantage of this power. When using computers as the main focal point in a system, duties that were once hardware generated have become more software dominated. The previous chapter outlined the aspects of approaching a design decision by providing a hardware function to answer the task. In this chapter, the hardware developed previously will be augmented by software. An attempt will also be made to determine when to use either in the design of a rover system. Therefore, when designing a control system using today's standards, a comprehensive trade-off study between the two is usually in order. Before embarking on such a survey, it helps to be literate of what hardware functions software can perform. To do this, let's examine the following examples.

CLOCK GENERATION

In traditional systems, a hardware oscillator coupled with some sort of frequency standard (crystal) would provide the

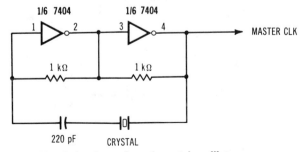

Fig. 2-1. Simple hardware-based crystal oscillator.

master clock upon which all other operations are based. An example of this oscillator is shown in Fig. 2-1. The output of this circuit is a square-wave signal changing at the rate equal to the frequency of the crystal. Using standard hardware this circuit contains numerous passive components, the crystal, and a hex inverter IC. There is an associated cost with these components, not only in the direct material prices, but also in the board space they consume and in the time required to test and debug such a circuit.

If you've only built electronic systems for your personal use, then only the direct material costs would matter. However, designing a unit for mass production requires the calculation of all three. Hardware-only systems are usually more parts intensive and, therefore, are more costly to produce. This brings us to the first comparison point in the trade-off study. Fig. 2-2 forms a matrix of functions versus concerns that can be used where performing any hardware-software analysis.

Being somewhat familiar with microprocessor electronics, I'm sure you're about to take issue with me over the fact that even microprocessors need the crystal, and some the whole circuit. Yes, you are absolutely right and in microprocessor systems the master clock should be in phase with this microprocessor's clock. The circuit in Fig. 2-1 is probably not directly replaceable. However, many other timing signals can be generated directly from the microprocessor through software. Let's look at another example. A system needs a 2-kHz, free-running clock and an output that occurs in sync with this clock. This secondary output is to pulse once every 16 beats of the master frequency.

Performing this function, in hardware, is indicated in Fig.

	PARTS COUNT	SPEED	DEVELOP COST	RECUR COST
HARDWARE	HIGH			
SOFTWARE	CAN BE SMALL			

Fig. 2-2. Framework of a software/hardware trade-off analysis chart.

2-3. Assuming a master clock frequency of 2 MHz, the 2-kHz clock would be derived from this through the use of a divide-by-1000 circuit. The three 7490 decade counters perform this portion of the duty. Next, a counter is needed to determine when 16 of these pulses have occurred. The 7493 and the AND gate perform this function. Together, you end up with five IC packages. Yes, they are all 14-pin ICs, and they can be packed into a small area, but what if you decide later that the second pulse should occur at 17 master clock periods? A major redesign of the circuit is necessary and worse if the master clock is to become something that is not evenly divisible.

A microprocessor implementation of this same function is shown in Fig. 2-4. This time, an Intel 8048 single-chip 8-bit microcomputer is used to perform all the functions listed before. Through instruction cycles, the 2-kHz master frequency is developed. The basic 8048 instruction cycle time is 2.5 microseconds. A 2-kHz clock has a cycle time of 500 microseconds; therefore, the 8048 must turn on the master clock output bit, wait the equivalent of 200 instruction times, then turn it off. This can go on indefinitely.

During those 200 idle instruction times, the processor may be doing other things. It also will be counting how many clock flip-flops it has made. When it reaches 16 or 17 or whichever you choose, it will also toggle the timing pulse output. Changing frequencies and pulse outputs only affects the instructions stored in the computer. In the case of an odd frequency, the microcomputer's crystal may have to be changed to effect a different instruction cycle time.

Comparing these two examples should graphically illustrate the gains a software approach can give. In this chapter we will go over the same applications shown in the previous one except we'll be doing it from the software system's point of view. Throughout our discussions, the trade-off matrix begun at Fig. 2-2 will be filled in and expanded upon.

CONTROL BY PLANNING

The purpose of the last few paragraphs was to attempt to convince you that software design can be as versatile as hardware. In microprocessor systems, often the software

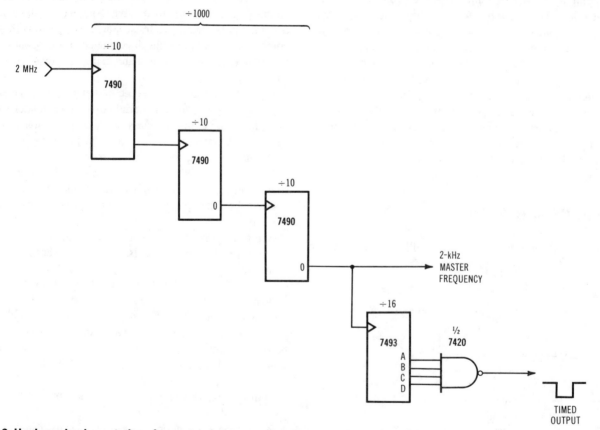

Fig. 2-3. Hardware implementation of a master clock generator with sync output.

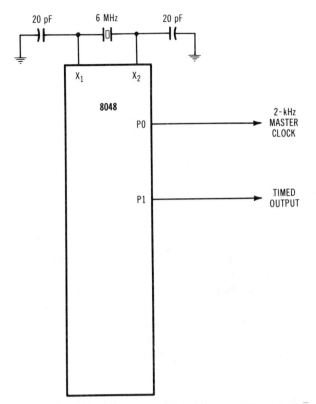

Fig. 2-4. Software/hardware implementation of Fig. 2-3 with 8048.

becomes the most critical item. It seems to be more labor intensive and requires much more planning. This is the subject of this section. A carefully planned software design usually turns out to be a properly working one too.

How do you plan a design using software? Well, remember the block diagram phase during the last chapter? There is a similar construct in the software world. In the last volume, I used the flowchart to indicate the actions of a program or subprogram. This same flowcharting method is used here.

For those of you unfamiliar with flowcharting, I offer a symbol key in Fig. 2-5 which illustrates the various block shapes and the meaning of each. You will see that I only use a few of the shapes in these diagrams.

However, before you get to the charting stage, many things must be decided upon. After going through the functionality lists on the system you are designing, make a detailed specification. From the specification, the hardware/software trade-off study must be undertaken. So far, the only block filled in on our chart is one of parts count. The next item we should look at is speed of execution.

In some instances, because the microprocessor has a finite instruction time, it may be impossible for it to perform a function. Take, for instance, the example of 2-kHz clock. If the requirements for a master frequency were a 200-kHz clock, then the chances of performing both the on/off to an output port and the counting to provide an output once every 16 times would be impossible for a microprocessor to meet given the 2.5 microsecond instruction cycle time. So speed is a factor and should be addressed as such. There are faster microcomputers, and they may have to be utilized in situations where hardware is not permitted and the software requires more performance. This completes a second item for the chart. It now appears as shown in Fig. 2-6.

Moving back over to the professional side of the house, in situations where the overall system being designed is to be producible and presumably field maintainable, both items figure prominently in the chart. The producibility issue brings up things like "time to design." In industry, time is money, and if it is going to take twice as long to design and build the first system, it will have to include features that allow trouble-free assembly, testing, and maintainability. In other words, things have to balance out. Hardware systems generally come out faster from an engineering team. Their actions are finite and usually fully spelled out at the beginning. Software, on the other hand, is very nebulous. In the early computer days,

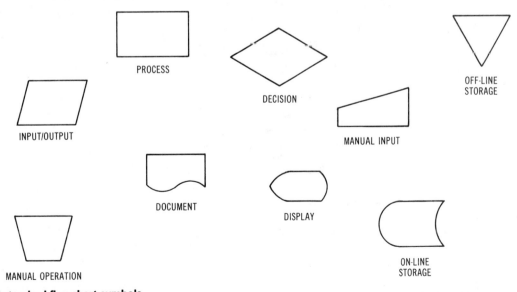

Fig. 2-5. Set of standard flowchart symbols.

software engineers were those "weird guys who spoke a different language, made their own hours (usually arriving late to work yet staying to 3 or 4 A.M.), and hated to use a pencil to write anything down." Today, most companies don't allow that.

Software documentation has come a long way and has further to go. The planning stages of a robot system had better include descriptions of each software-performed task. These descriptions should be as detailed as a theory of operation for hardware. If the microprocessor is to toggle an output port bit at 2 kHz, the description should list the port arrangement, the assignment, and the method of counting the time between toggles. This type of planning makes the design task simple. A high-level programmer, after writing a detailed description, can hand it off to a junior partner for coding. In this way the higher-level person is free for the planning of the next job and the junior programmer has the chance to learn a well-thoughtout system by coding the examples. Sounds like it makes the most sense, yet most companies are still not performing it that way.

The last item for the chart would be overall project cost. I know I've already mentioned cost throughout all the items, but it is important to list it as a separate entity at the bottom of the list. The filled-in items above each carry some monetary value, be they plus or minus. The completed chart is shown in Fig. 2-7.

When the functions, responsibilities and detailed plans are prepared, flowcharts may be utilized to help design the coding structure. A system diagram, very much like a hardware block diagram, may be included to help visualize the entire system structure. The main program loop and all subroutines should be shown in their hierarchical order on the chart. An example of such a software system diagram is shown in Fig. 2-8.

Of course, through all this, the actual design function becomes the end product that we are striving to attain. The rest of this chapter will be devoted to implementing various operational functions with an equal component of both hardware and software. Let's go into some specific duties as they apply in rover design.

MOTOR CONTROL VIA SOFTWARE

As we mentioned in the last chapter, the most outstanding of features in a robot rover is the drive system. Without it the robot does not "rove." In that section, we also went over a few methods for controlling their operation and direction in conjunction with various guidance methods. Here, we will leave the guidance for later and just concentrate on making those wheels go round.

Let's assume that a typical rover system is composed of two drive wheels. You have been over their operation several

	PARTS COUNT	SPEED	DEVELOP COST	RECUR COST
HARDWARE	HIGH	VERY FAST		
SOFTWARE	CAN BE SMALL	SLOW		

Fig. 2-6. Trade-off analysis chart with speed comparison.

	PARTS COUNT	SPEED	DEVELOP COST	RECUR COST
HARDWARE	HIGH	VERY FAST	LOW TO MED	MED TO HIGH
SOFTWARE	CAN BE SMALL	SLOW	VERY HIGH	LOW

Fig. 2-7. Completed software/hardware trade-off analysis chart.

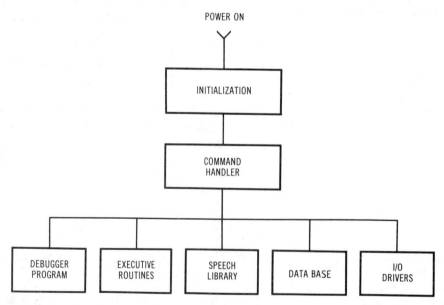

Fig. 2-8. Typical software system diagram.

times and understand how to make the vehicle go forward or reverse and how to make it turn. Wheels and the drive motors that move them are driven in one direction by the way current is flowing. Reversing the current will reverse the direction of the wheel. You have seen this done with the use of the double-pole relay. Fig. 2-9 reiterates the typical connection scheme of a two-motor drive platform.

As the two relays rest (remain unenergized), they supply an opposite direction of current to each because each motor is physically the opposite of the other. The motors will turn producing a straight line motion (either forward or backward). Notice that each relay may be energized separately. Therefore, either motor may be reversed on demand without affecting the other. This is where the software comes in.

Suppose that the relays and motors happen to be connected in the diagram to supply a constant forward motion. Each relay would be connected to an IC driver that will handle the surge of current necessary to energize it. Fig. 2-10 depicts all this plus a connection to a standard parallel I/O microprocessor peripheral chip called a *peripheral interface adapter* (PIA). This part allows a standard microprocessor, one without any on-board I/O, to control various inputs and outputs directly, just like the 8048. You may find this application more appealing as it might apply to your particular situation. All these examples may be simulated using microcomputers. There is no need for a stand-alone system when researching an idea.

Back to the diagram. One of the PIA's output lines is connected to each of the relay drivers. Now, through software, either relay may be reversed. Adding a third relay and driver, as shown in the diagram, allows the entire vehicle to stop as well as turn. Let's investigate the program actions to do all four motion functions and the function of enable/disable.

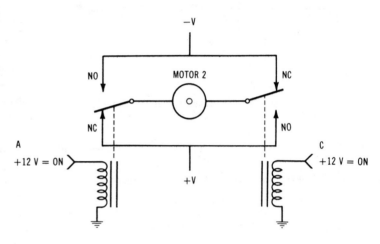

Fig. 2-9. Schematic of a two-motor mechanical drive system.

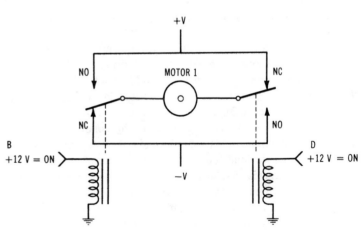

A	B	C	D	DIRECTION OF ROVER
0	0	0	0	FORWARD
1	0	1	0	LEFT
0	1	0	1	RIGHT
1	1	1	1	REVERSE

1 = ON   0 = OFF

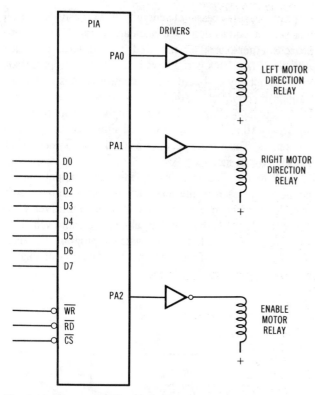

Fig. 2-10. Microprocessor-interfaced two-motor drive control.

The example shown in Fig. 2-11 shows the steps necessary for a microprocessor to turn the vehicle to the left. At the beginning of the program, the motor enable relay is assumed in the off condition (output 3 = 0). The left motor drive bit (output 1) is set high (1) to reverse direction of the wheel. Then, the right motor drive bit (output 2) is reset to a logic zero, which will force it to remain running forward. At the end of the program, the motor enable bit is set to a logic 1, which turns both motors on. The vehicle will then turn left.

A somewhat slow implementation, yet quite effective and more general purpose, is shown in listing form in Fig. 2-12. This is a motor control program written in Microsoft BASIC. It assumes the 8255A PIA chip is located at addresses 49152 (port A), 49153 (port B), and the control port is located at address 49154. The PIA is configured using this control port. For this example, port A is used only, and it is configured as all outputs. The decimal code to write to the control port to configure this way is 159.

Follow the program flowchart in Fig. 2-13 to learn about what this program does. First, it intitialzes the control port on the PIA. This done, it immediately turns off all relay drivers. This safe condition allows you, the programmer, to begin at a known state. Why did I say "you, the programmer"? Well, now the program asks you to select a direction. After you choose, it will perform the action through the methods we have been discussing. Turning on and off port bits for control of a system is called *control by bit manipulation*. The flowchart does not show the separate direction subroutines; they are shown in Fig. 2-14.

There is nothing magical about this method of control. You are simply implementing a hard-wired control function with software. If you own a computer with joysticks, the program can be modified to incorporate them as the operator controls instead of the cumbersome keyboard. The changes shown in Fig. 2-15 are specific for the TRS-80®* Color Computer. Your particular system may handle joystick inputs differently. In the TRS-80® they represent digital values from 0 to 64 in each of two directions (X,Y). I have arbitrarily set 40-20 as the center zone where all motion stops. Increasing the Y direction number will move it to reverse while a decreasing number will switch it to forward. The X direction value will similarly turn left on an increasing value, right when decreasing.

ADDING A STEERING WHEEL

There are many applications where, for one reason or another, the two-wheel drive system will not fit the bill. Three- and four-wheel vehicles can be made to be controlled via software almost as easily as the previous one. I say "almost" because you have the matter of the steering wheel or wheels to deal with. Yes, a turn right can be accomplished by moving the steering wheels hard right (to where they can't turn any more), but in most cases this type of turn will result in a circular move rather than the imitation of a tangent to a circle.

The key here is proportionality and how to achieve it with software. To begin to even think of controlling the direction of a wheel proportionate to some reference, we must investigate certain methods of determining where the wheel is at any

---

*TRS-80 is a registered trademark of Radio Shack, a division of Tandy Corp.

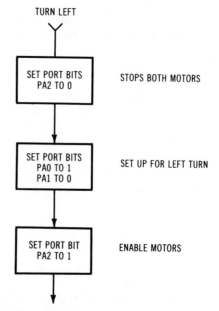

Fig. 2-11. Flowchart showing action of processor to command motors connected, as in Fig. 2-10, to turn left.

```
10 REM FOUR DIRECTION
15 REM MOTOR CONTROL
20 REM PROGRAM
25 REM c 1983 M. J. ROBILLARD
30 REM
35 REM INITIALIZE PIA
36 REM
40 POKE 49155,159
41 REM
45 REM INITIALIZE MOTOR STATUS
46 REM
50 POKE 49152,0
51 REM
55 REM OPERATOR COMMAND HANDLER
56 REM
57 CLS
60 A$=INKEY$
65 IF A$="" THEN GOTO 60
70 IF ASC(A$)=94 THEN GOSUB 200
75 IF ASC(A$)=10 THEN GOSUB 300
80 IF ASC(A$)=8 THEN GOSUB 400
85 IF ASC(A$)=9 THEN GOSUB 500
90 IF ASC(A$)=13 THEN GOSUB 600
95 GOTO 60
```

(A) Main program.

```
97  REM CONTINUATION OF MOTOR
98  REM   CONTROL PROGRAM
99  REM
200 REM FORWARD ROUTINE
201 REM
203 CLS
205 PRINT"FORWARD"
210 POKE 49152,0
215 POKE 49152,7
220 RETURN
221 REM
300 REM REVERSE ROUTINE
301 REM
303 CLS
305 PRINT"REVERSE"
310 POKE 49152,0
315 POKE 49152,4
320 RETURN
321 REM
400 REM LEFT TURN ROUTINE
401 REM
403 CLS
405 PRINT"LEFT TURN"
410 POKE 49152,0
415 POKE 49152,6
420 RETURN
421 REM
500 REM RIGHT TURN ROUTINE
501 REM
503 CLS
505 PRINT"RIGHT TURN"
510 POKE 49152,0
515 POKE 49152,5
520 RETURN
521 REM
600 REM STOP ROUTINE
601 REM
603 CLS
605 POKE 49152,0
610 PRINT"VEHICLE STOPPED"
615 PRINT
620 PRINT"SELECT THE FOLLOWING"
625 PRINT "1) RESUME PROGRAM"
630 PRINT "2) EXIT TO BASIC"
635 A$=INKEY$
640 IF A$="" THEN GOTO 635
645 IF A$="1" THEN CLS:RETURN
650 IF A$="2" THEN CLS:END
655 GOTO 635
660 REM
```

(B) Subroutines.

Fig. 2-12. BASIC implementation of a simple four-direction motor control program.

given time. In Volume 1 of this series is presented a whole chapter on the design and use of position sensing transducers. Here, we will apply that knowledge in a software-controlled proportional steering system.

There are four methods of physically determining the position of our steering wheel in a robot rover. Fig. 2-16 depicts them, as I see them. The first is similar to the method used to display the current position of the playback head in an eight-track car stereo player. A series of contacts are placed around the perimeter of the steering shaft, perpendicular to the wheel. A contact is affixed to the center shaft. Whenever the wheel moves, so follows the center contact. As it rotate it contacts each of the position sensors in order, one by on microprocessor connected to such a system is shown

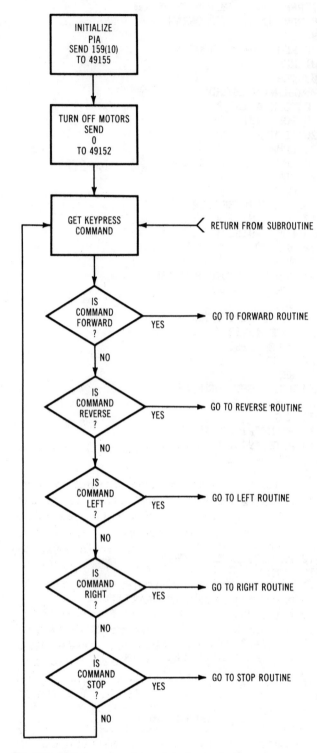

Fig. 2-13. Main program loop of motor control program.

can determine angular movement, and through the use of software, report on the absolute position of the wheel. I introduce the words *angular* and *absolute* here, because in this system there is only one detector (contact). There are numerous position marks on a circular plate that rotates along with the steering shaft that can be used to determine position.

Examine this implementation in Fig. 2-18. The flowchart comes later. For now, let's concentrate on the hardware interface. It sure is simple. Every optical mark or magnetic spot on the disk will pulse into a PIA input. Each degree of turn is then reported. But how do you know where you are initially? Well, do you see the second detector? That is there to report on a "sync" mark on the disk. This special section will tell the microprocessor that the disk (and the wheel, for that matter) is at its center position.

Now let's see what the software does with these pulses. Fig. 2-19 outlines the flow of the program. Initially a location in memory is cleared. This location is called the position register. It is here that the known wheel position will be loaded. After clearing it, the steering wheel sync detector is read. If the wheel is not centered, then the command to move in a clockwise direction is ordered. When the sync area is found, the wheel will be stopped.

From that point on, any deliberate movement of the wheel to the right will result in the position register being incremented by the action of the degree pulse detectors. When the wheel is moved left, the register is decreased. A fairly accurate description of the whereabouts of the wheel, at any time, can be accomplished using this technique.

To turn a given amount, simply load the absolute position of where you want the wheel to turn to into another location called the *compare register*. If the value there is greater than the current position register, a command is issued to turn right until the two match. The opposite is effected when a lesser number occupies the compare register. The flowchart of Fig. 2-20 reiterates these actions for clarity.

The last method to be discussed is to emulate the action of a servomechanism. Here, a variable resistor is connected to the steering shaft. As the wheel turns, the resistance changes. In this system, the center position must be known beforehand and programmed into the computer. The position of the potentiometer (variable resistor) is measured by converting the resistance value into a varying voltage. This voltage is then applied to an analog-to-digital (A/D) converter (Fig. 2-21).

As mentioned, the center position of the wheel will produce a known digital value. Upon initialization, the program will read the value present at the A/D converter. If it is lower than the known center, the wheel is turned right until the values match. A higher value will result in the opposite. Sound familiar? It should. This is the analog equivalent to what was done before with optical or magnetic spots. The flowchart of Fig. 2-22 details the operation specific to the analog method.

From here, we depart from on and off, left and right.

2-17. Here, as the figure shows, the logic ground applied to the center shaft is read through the contacts when it is located on one. The eight contacts are read as 1 byte. The one that shows up as a logic low is the current position.

The second and third methods are similar, and they will be discussed as one. An optical or magnetic detection system

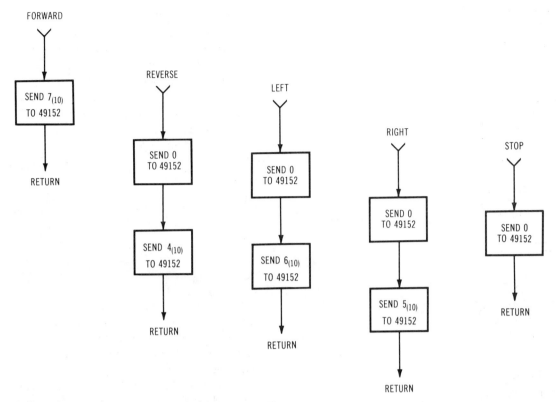

Fig. 2-14. Subprogram flow diagram of motor control program.

Although these actions are the most critical in the design of rover systems, there are other niceties.

SPEED CONTROL

The function of speed control of motors usually involves a great deal of hardware. In fact, some of the methods that will be described involve quite a bit for a software chapter. These controls are not necessary in a majority of cases. You will find that three speeds may be plenty. I have seldom needed more than two. All this aside, let's get into some of the more orthodox methods of keeping pace with the drive motors.

The simplest method of controlling the speed of the vehicle is to select a finite set of speed ranges. Each range would correspond to a given motor voltage. As the voltage is decreased to a motor, the slower it will turn. A circuit that performs this function is shown in Fig. 2-23. Once again, there is an overall enable/disable control bit and two direction bits. From here, it is all new. There are three speed-control bits that energize the appropriate relay to produce the desired rate of travel. In all only 6 bits are necessary to control the circuit. A flowchart depicting the computer's actions, given the speed rate, is provided in Fig. 2-24.

There are several problems with this type of speed control. The first is that some motors do not perform equally as well at voltages other than those specified for it. Another source of trouble is that the speed will not be constant. The vehicle will tend to travel faster when traveling down hills and slower when traversing inclines. A true speed control needs some sort of feedback from the wheels to help regulate the rate of travel.

When you drive a car manually, the speedometer performs the function of rate feedback. Your brain has set a value of speed to reach, and the difference between that speed and the one showing on the dashboard can be classified as the error value. This value may be positive or negative. The microcomputer has the ability to regulate speed in the same way the operator of an automobile does, provided it has similar feedback mechanisms and rate control linkages.

In the microcomputer system, this control linkage can be a varying voltage as long as it varies accordingly with the speed the vehicle is traveling. Feedback can be provided by attaching a tachometer or similar device to the drive wheel itself. Industrial tachometers produce a voltage proportional to the speed at which they are driven. Fig. 2-25 illustrates the mechanical use of the tach in a rover vehicle.

Another type of speed-sensing device uses an optical disk much like the steering wheel position sensor discussed earlier. Remembering the operation of this device, a pulse will result each time a hole is detected. When the disk is connected to the drive wheel shaft, pulses will occur frequently during travel. Depending on the distance between the slots and the speed at which the disk is rotating, this frequency can be used to determine speed.

The trick is to measure the time span between holes detected. This time will give the period of the frequency. If it is not faster than the microcomputer can measure (highly

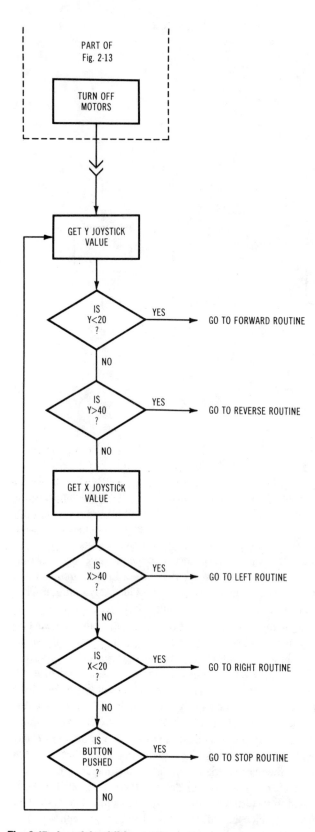

Fig. 2-15. Joystick addition to the motor control program.

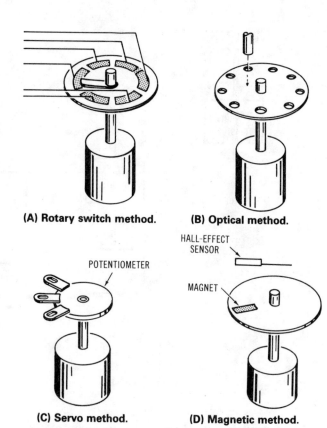

(A) Rotary switch method.

(B) Optical method.

(C) Servo method.

(D) Magnetic method.

Fig. 2-16. Four possible methods of providing feedback of position from a steering wheel.

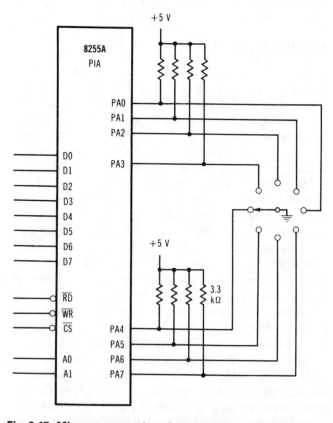

Fig. 2-17. Microprocessor-interfaced eight-position steering wheel position transducer.

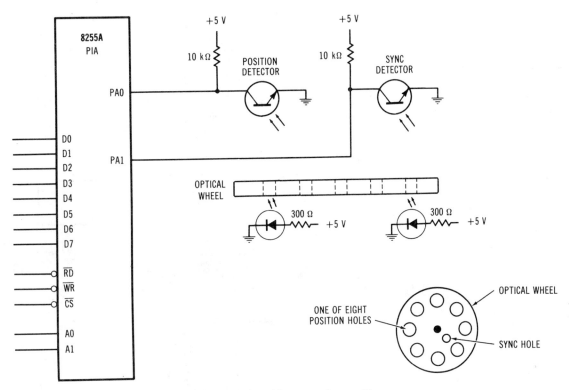

Fig. 2-18. Microprocessor-interfaced eight-position optical position transducer with sync.

unlikely), then this will result in an accurate speed measurement system. Let me be a little more specific.

Pulses coming from the detector are connected into the microcomputer circuitry through an I/O port. When a logic low is detected, the program begins marking time until the next low pulse is encountered. The time span value that is captured can be compared against a set value. If the read value represents a slower speed than desired, the program will connect the motor to a higher voltage. When the value read is faster than the desired speed, a lower voltage is used. The only difference between automatic speed control and the circuit of Fig. 2-23 is the addition of an optical sensor.

Fig. 2-26 is a flowchart of the operation of a program that governs speed through the use of the optical transducer. This circuit, coupled with the steering mechanisms discussed and a dedicated microprocessor, will yield intelligent motor control. Has software simplified the design of such systems? No, it probably has not; however, there are areas where high current drive is not necessary, and you will see the power of programming.

The next area covered in the last chapter was that of guidance. In rover systems, this task goes hand in hand with motor control. You have seen the circuitry involved in one-, two- and even three-detector optical track systems. Let's investigate ways that software can alleviate the rover of some hardware. The first type of tracks we should look at is the single-detector system.

Remember that it was very hard to design an optical track circuit with only one detector. Leaving the path is announced by the detector; however, the vehicle has no physical way of telling in which direction (left or right) it is wandering. A microprocessor, on the other hand, has much more flexibility. Assume this same microprocessor is controlling the motors. When the detector signals a loss of guide path, the processor can experiment with direction by first stopping the wheels, then turning a little left and checking the detector. If this fails, a slight turn to the right may find the path. Granted, the vehicle would look a little strange groping around each time a bend comes up in the path, but the application is explained here to prove a point. Look at the schematic in Fig. 2-27 and compare this to the hardware equivalent back in Fig. 1-12. Sure the microprocessor will add circuitry, but chances are the original circuit would reside in a vehicle equipped with a microprocessor for other reasons.

A much more practical system is shown in Fig. 2-28. Here, two detectors are used to provide a guidance mechanism that easily keeps on track. The only added circuitry is the comparator to support the extra detector. These comparator ICs come with four individual circuits in them so the package count remains the same.

The program to support this scheme would first assume it has been placed over the track. It would then read both detectors. If both were showing that they are over the pathway, then the vehicle is free to move forward. During motion, the detectors are read at regular intervals. When and if a difference occurs, the processor will do one of two things. If the left detector is off the track, the right drive wheel is reversed, which will pull the vehicle more toward the path.

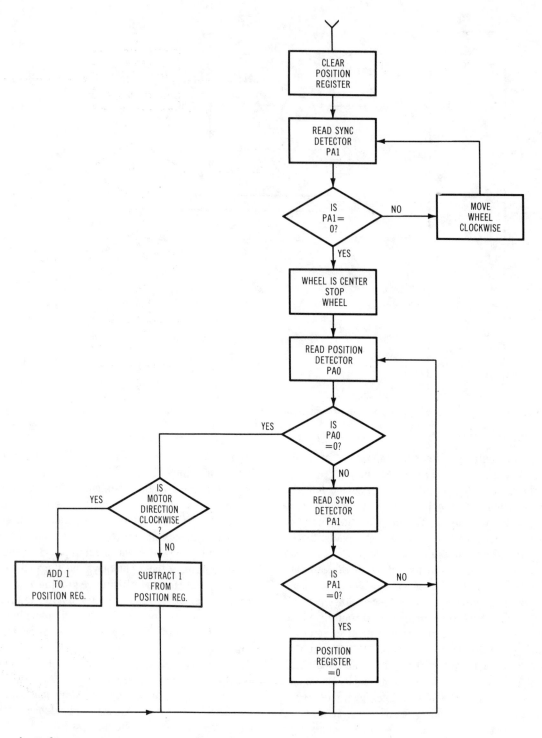

Fig. 2-19. Flowchart of program necessary to control transducer shown in Fig. 2-18.

The opposite is the case if the right detector is showing itself as off-path.

The flowchart in Fig. 2-29 details the program for the two-detector system. Using an arrangement such as this will allow you to experiment with all the different combinations of straddling a guidepath without ever changing circuitry. A few adjustments to the program allow the detectors to react when they see the guidepath instead of when it's not there. You can even design an algorithm to find a guidepath when the vehicle is started from a point far away from one.

Let's examine the mailmobile application to get a feel for the level of software support necessary for such a project. In this study, the guidance and motor drive algorithms will surface and lead to other areas of concern.

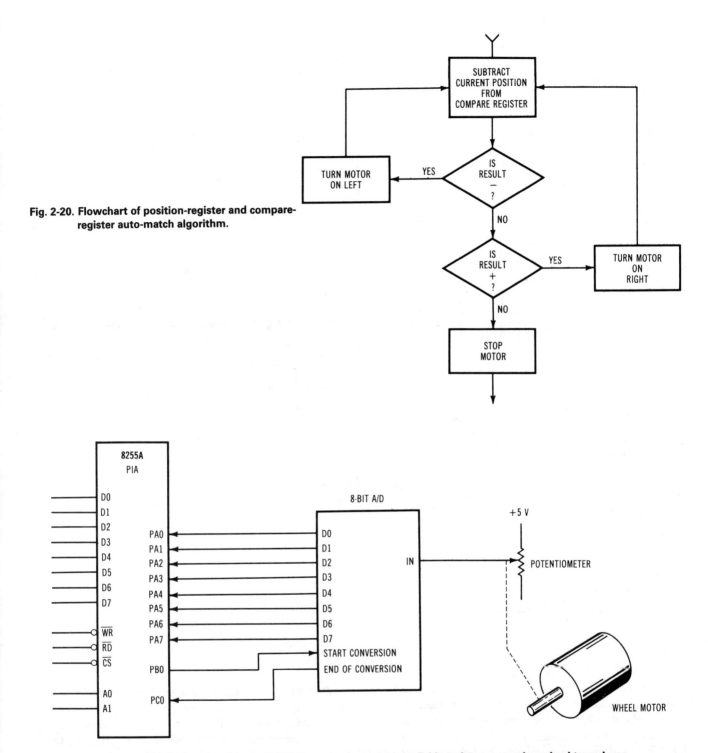

Fig. 2-20. Flowchart of position-register and compare-register auto-match algorithm.

Fig. 2-21. Microprocessor-interfaced analog-to-digital converter to measure variable resistance steering wheel transducer.

MAILMOBILE SOFTWARE

If you recall, the specifications listed for this vehicle included four-wheel drive, optical path guidance, and route stop-point recognition. Let's assume that the mobile is garaged in the mailroom overnight where its batteries are charged for a full day's work. When it is pressed into service, the mailroom personnel have already loaded it up with the day's first deliveries. It is loaded into a freight elevator so that forward motion will carry it out the doors when they open. The mailperson selects a floor on the elevator panel and closes the door. Up, up, up the robot travels until the elevator reaches the desired floor.

Problem 1: How does the robot know the elevator doors have opened?

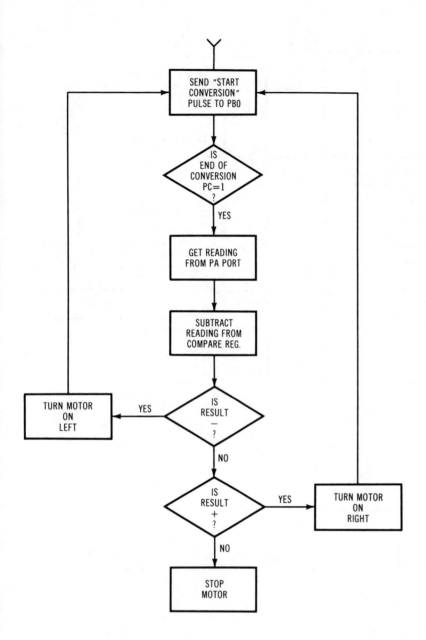

Fig. 2-22. Flowchart of a servo-based transducer system.

Obviously, some sort of communication must exist between the elevator and the robot. A simple system of the detection of light when the doors open could solve the problem. That gives us our first design goal. The software must, when initialized, assume it is going from a light area to a dark one (elevator) then, when the light appears again, it is time to move forward. Fig. 2-30 shows a flowchart of that logic. Now, every time the robot is initialized or reset, it will go through this process.

We could even expand on this to include the ability of the robot driving itself into the elevator after initialization. In this way, the robot would assume the doors are open (something the operator must set up first), and then it moves backward until its rear bumper contacts the rear of the elevator. When it does, it will then go to the light-dark-light algorithm. Incorporating this feature will make it easier for the mail personnel. All they have to do is load up the vehicle, open the doors to the elevator, and start the robot. The rest is automatic.

If we have settled on this as our auto-start program, then use the complete algorithm flowchart in Fig. 2-31. The next encounter happens when the elevator doors open and the light goes on. Where do we go from here? We will have to assume that the guidepath has been painted right up to the door of the elevator. If this is the case, the robot throws itself into forward gear and moves on out of the elevator. It then finds the guidepath and continues along. Fig. 2-32 illustrates our next problem.

Problem 2: At a fork in the guidepath, which way do I go?

Actually, this is where I wanted to lead you. As you can see, simple track guidance methods are not enough for intel-

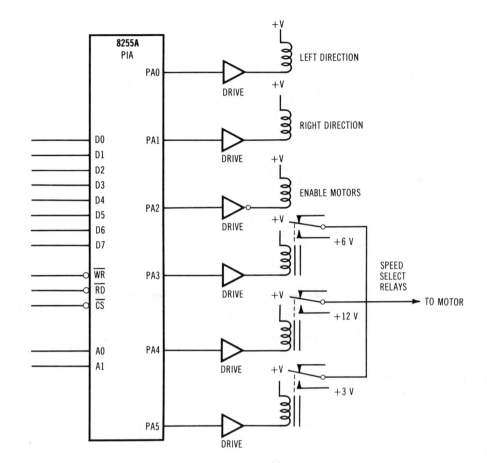

Fig. 2-23. Simple three-speed voltage-controlled speed controller.

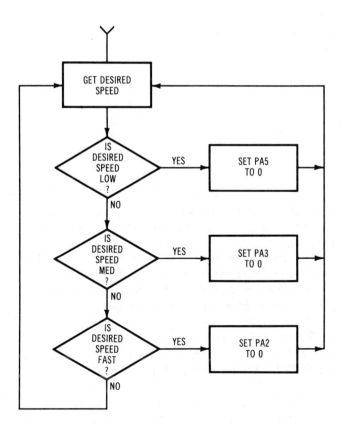

Fig. 2-24. Flowchart of program used to control Fig. 2-23.

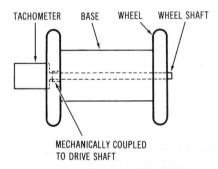

Fig. 2-25. Diagram showing use of a tachometer to provide speed feedback.

ligent motion. The reason there is a fork in the path is because this other side is the return way back to the elevator. There needs to be something to tell the robot, at the next fork, turn right. A separate detector may be added that looks for commands painted as a series of dashes on the floor. In this case, the command track could be located to the right of the guidepath.

In a mailmobile application, I can't imagine too many of the special commands would be necessary. Let's list what information might be passed through this method of commu-

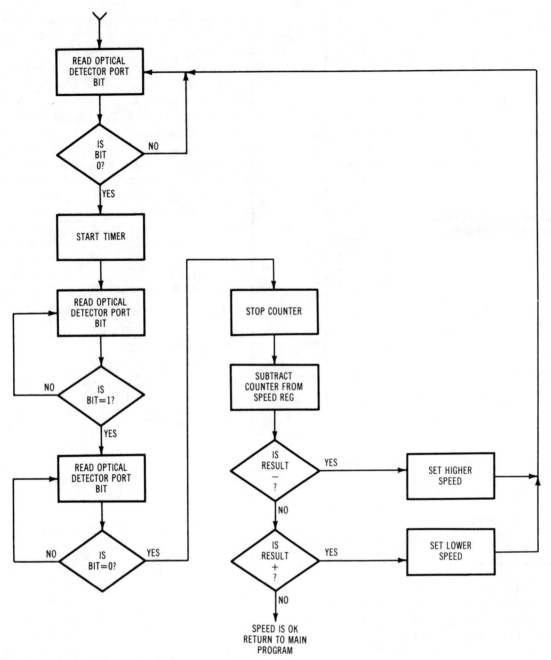

Fig. 2-26. Flowchart of a speed control program using an optical sensor for rate.

nication. First, you will need turn right at fork and left at fork, which gives us two commands. Another is a planned mail-drop point. From there, the only other one we could add would be, possibly, a stop and await elevator or assume elevator is waiting and go into initialization routine. Altogether, that makes four.

Four commands can be passed to a single detector through a time-serial system. In this system, a command sync stripe is detected, and then up to four additional stripes are read depending on the command. A time-out must be designed so that the robot doesn't continuously look for additional stripes when only one is encountered. Maybe an arbitrary number could be 5 seconds. So, if, after 5 seconds, no other command stripe is encountered then the command is assumed to be received and is acted upon. An algorithm to perform this detection is included in Fig. 2-33.

Let's go through the flowchart. Initially, the detector will see nothing. When a detection has been made, it is assumed to be a sync bar. From there, a counter is cleared and readied to receive the number of following bars. Each time a new detection is made, the counter is incremented. Between these detections, the 5-second time-out timer is initialized. If the time-out happens, the number in the counter is used to select a command. That command is then acted upon.

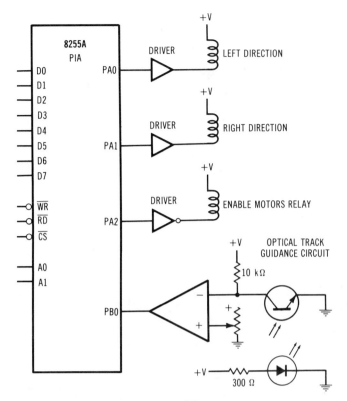

Fig. 2-27. Microprocessor-interfaced single-detector guidepath circuit.

Obviously, in a four-command system, this method will work well. However, can you imagine the job of the path painter in a system with 200 commands? Surely, you would use another method of communication. One of the methods in use today involves standard bar codes much like the ones printed on the side of grocery store food items. In our limited system, though, we don't need quite this sophistication.

Well, let's look at what we've covered. The mail robot has been programmed to go onto the elevator, get off at its stop, turn right or left at a fork, stop at selected mail drops, then go back into the elevator for the return trip to the mailroom. We haven't said much about those mail-drop stops. After all, they are the reason the robot exists in the first place.

Whenever the stop code is encountered, the mobile should beep some sort of annunciator and stop for some period of time. Realistically, after a beep you should give a person at least 30 seconds to reach the vehicle. One beep may be unnoticed, so I propose the following. The mailmobile receives the stop code. It then gives two short beeps and waits 20 seconds. If the mail in the slot is not taken it gives another beep. After the total 30 seconds have elapsed, the vehicle moves on its way. Recall that I mentioned there are often "stop bars" on these vehicles that may be manually operated to extend the time period.

This brings us to the flowchart in Fig. 2-34. Here, you can see that, upon reception of the stop code, the beep occurs twice. The vehicle stops and the 20-second timer is initiated. An optical detector in the bin housing the mail for that stop will then be interrogated to see if the mail has been retrieved. If it has, then a status flag is set. Upon reaching the 20-second limit, the status flag is tested. If it is set, another beep is sounded, then it goes on to the final 10-second timer. At the end of the total 30 seconds, it will move on.

Where is the stop bar in the program, and why haven't the front and rear collision bumpers been included? This brings up a good point. These three items may be considered out of the ordinary occurrences that will interrupt the normal flow of the program. In that way, we could include an interrupt structure and service routine that will react to one of them being activated. Let's discuss the design of such a program. Fig. 2-35 is a flowchart of the interrupt routine.

Looking at the figure, we see that when an interrupt occurs, there must be some hardware-based method of determining where it came from. Let's assume this is done and fed into the processor through a 2-bit port. Upon interrupt, the robot will read this port. Depending on the code it receives, it will either stop (collision of front or rear bumper) or extend

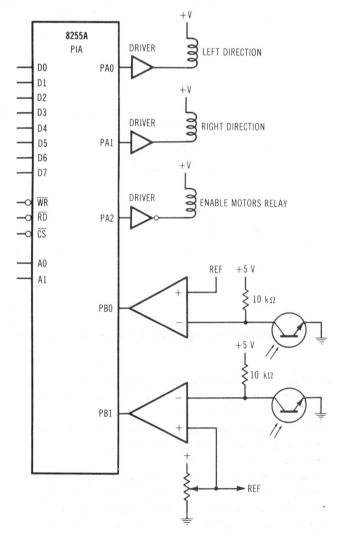

Fig. 2-28. Microprocessor-interfaced two-detector optical track guidance system.

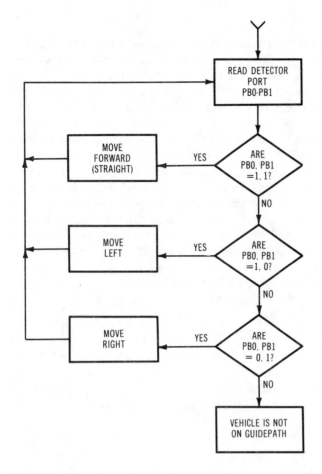

Fig. 2-29. Flowchart of two-detector software-controlled system.

Fig. 2-30. Flowchart of initial actions of a mail-carrier robot.

the mail-drop wait period (stop bar hit). I have added a fourth input called "low battery." When this occurs, the robot will stop and beep its annunciator once every 10 seconds. The interrupt routines will return to where they were called from, with the exception of the low-battery code, which will simply loop there until maintenance people arrive.

So far, we have been dealing in the theoretical. I would like to briefly describe an implementation of the previously described system using a single-chip microprocessor. Whenever real hardware comes into play, some of the theory usually suffers. In terms of the mailmobile, however, you will see exactly what we have discussed. The schematic of the system is shown in Fig. 2-36. I appologize for mixing so much hardware with the software in this chapter, yet I find it extremely difficult to talk about one without the other. Today's engineers are becoming more or less "multiware" engineers because of this.

Going over the schematic, the 8748 in the middle is the microcomputer. The mail robot's control program is stored in the EPROM located on-chip. All the detectors, motors, bumpers, and steering position indicators go through the ports on this device. The program algorithms we discussed so far can all be implemented within this device. Let's block out some structure of the internal ROM.

Upon power-up, the processor will go to location 0 and there will pick up a jump instruction to the main command input loop (Fig. 2-37). Here, it is basically checking the status of the five manual-control buttons that help the mail personnel maneuver the robot into position. As you can see from the flowchart, as each direction button is pressed, the vehicle will move until the button is released. In this way, personnel may "drive" the robot around to various stations where mail may be loaded onto the carrier trays. The fifth button is the *start* command that will send it into the go-on-the-elevator routine that we discussed before.

Electrically, these buttons may be remotely controlled via a cable so that they could be held by the operator as a calculator-sized pendant. A sixth one could be added to stop the program. If you really get ambitious, an infrared wireless link may be designed so that there would be no cumbersome wire around to get hung up on things around the mailroom.

Back to the software, the main loop will remain in effect as long as the start button is not pressed. As soon as it is pressed, the routine jumps to the path track mode. The flowchart of this part of the software was shown in Fig. 2-31. The program goes on following the diagrams of Fig. 2-29. This is the

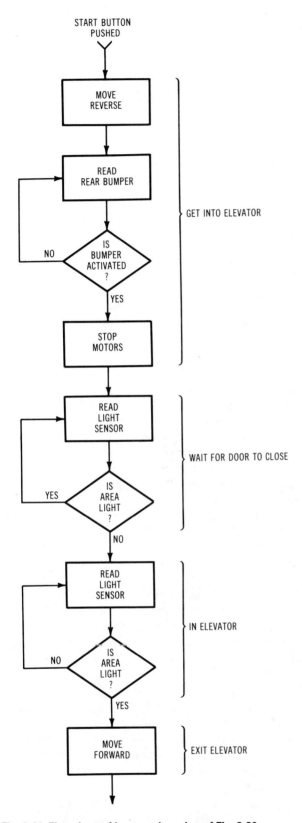

Fig. 2-31. Flowchart of improved version of Fig. 2-30.

routine that guides the vehicle down the halls through the use of its two optical path detectors. There is a change here though, where the flow brings you back up looking at the path detectors again. This is where the command detector flowchart should go. The revised chart is shown in Fig. 2-38.

As you can see from here, the program will stay in the path track mode, maneuvering the vehicle along the halls and reading any commands that may come along. The only other portion of the software not included is the interrupt routine. We went over these actions recently, blocked them out in the flowchart in Fig. 2-35, and mentioned that they would be interrupt routines. The only thing to add is that in an 8048 system the interrupt line (pin 6), when activated, will propel the program to location 003 in the processor's control program memory. Here, a JUMP instruction will be placed to send the flow to the interrupt routine. A complete flowchart of the mailmobile 8048 application is shown in Fig. 2-39.

DISTRIBUTED SOFTWARE

In the last chapter, I harped on the values of distributed versus single processor control. So far, in this chapter, all I've been doing is devoting time to the latter. I've decided to cover the subject from the standpoint of providing information necessary to program a master processor, which primarily talks to the Mostek SCU20 slave I/O chips. I will go over all the commands available and touch on some possible applications of the part in rover design.

Communication from the master processor (host) to an SCU20 is done over a serial link. A message is composed of a minimum of four characters. Each character has the following characteristics:

1 Start bit (always low)
8 Data bits
1 Even parity bit
1 Stop bit (always high)

These characters are sent out at one of four baud rates. The actual rate, be it 300, 1200, 2400 or 9600, is determined by different strapping options external to the SCU20 part. The host, naturally, must communicate at the same rate as the slaves are strapped for. The host will initiate all communication on the link. The message sent from the host to an SCU20 is called *command message*. When the host sends a command message to a slave, that particular SCU20 will perform the action requested, then return a *response message* back to the host. Response messages are only generated in conjunction with a previously sent command message.

All command messages conform to a specified format. Each character in the message carries a certain piece of information. Fig. 2-40 illustrates the format used. The first character transmitted by the host, when commanding a slave node, is the command header byte. This value (hex 01) will tell all the slaves on the link that a command is coming down the line. At this point, all slaves are listening and decoding the message. The next character is the binary code (00-FE) of the address of the slave it wants to communicate with.

Each SCU will now decode the address character and

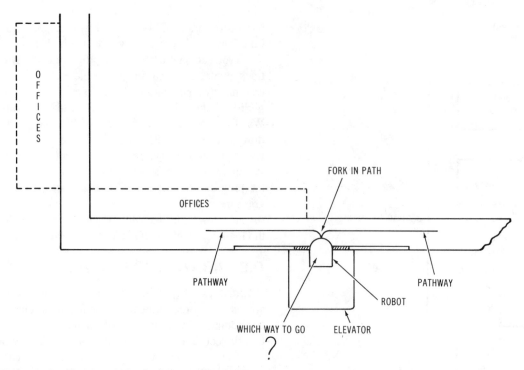

Fig. 2-32. Diagram showing a need for decision-making logic as soon as the rover leaves the elevator.

compare it with the externally strapped address it possesses. If there is no match, the SCU will not take any action but to wait for another command header byte to come along. It will ignore all other codes.

The SCU that does possess the correct address will listen further to the next character, which is the command code itself. This byte tells the slave what to do. We will be going over all the available command codes later. For now, let's stick to the general format of the message. Following the command code will possibly be some data that supports the particular command. This data consists of a series of characters. To top off the message an error-detection code called a *longitudinal redundancy check* (LRC) character is included (more on this later).

The response message is very similar in structure to the command message. Fig. 2-41 illustrates the format. As you can see, the first character is called the response header. Its value is a hex 02, which indicates to all on the link that a slave is responding. After that, comes the address of the responding node. This will match the address of the previous command message. If it doesn't, then there is an error on the link somewhere. Next will be an echo of the command code, which is now called the response code. There is one exception though, if the SCU20 addressed is not ready, needs to be reset, or finds a problem with the command message, it will respond with a special *NAK* character in the response code location that indicates an error.

After the response code, any data characters necessary to support the response code are transmitted, and the last character is the longitudinal redundancy check code. Together, the command message and the response message comprise the total message system used over the SCU20 link. This system may be referred to as the *protocol* of the communications link. As with any communications line, errors are possible, and if they remain undetected, they may cause serious difficulties. In a mechanical rover system, you don't want a communications error responsible for the rover careening down a flight of stairs or ending up in the lunchroom terrorizing its patrons. Errors such as these can be weeded out through various techniques. The type used in this system is called the *LRC method*.

Remember the longitudinal redundancy check byte at the end of each message? Well, that's what LRC stands for, and it is a very involved method of assuring the integrity of the data being sent. Basically, each data bit of the LRC byte represents that same bit position in all bytes of the message. Bit D0, in the LRC character, will contain a logic 1 if the total number of D0 bits in the message, containing a logic 1, is odd. Bit position D1 will exhibit the same for all D1 positions and so forth. In the end, you have an 8-bit character that represents an error-detection code. Your host has calculated the LRC of the transmitting command code and sent it out. The receiving slave will also calculate it and compare it against the one being sent from the host. If the two match, the message is allowed. When they don't, the NAK message is sent. The same follows for the response message. The slave will calculate and send the LRC, and the host will calculate and check the received one. In the case of a response message with an error, the host would reissue the command.

Speaking of commands, let's get into the available commands for the slave SCU20s. Remember, from the description of the last chapter, that there are three general-purpose

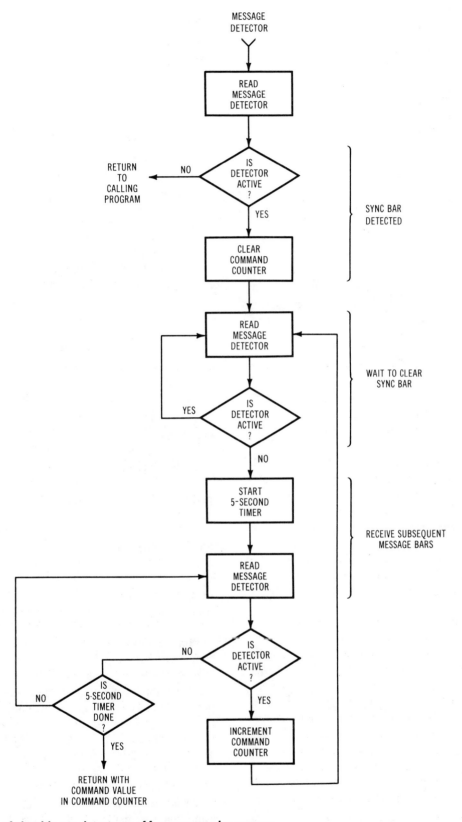

Fig. 2-33. Flowchart of algorithm to detect one of four command messages.

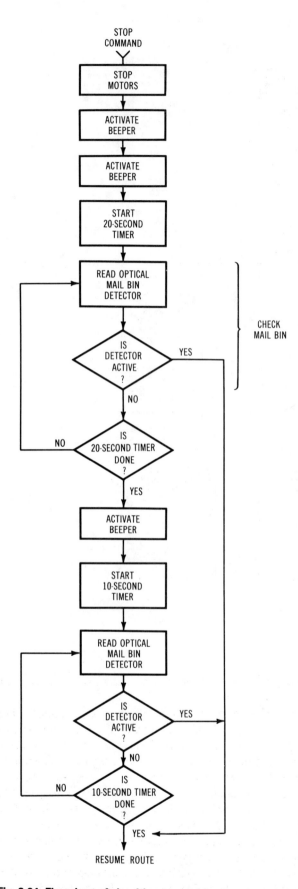

Fig. 2-34. Flowchart of algorithm to provide stop annunciation.

8-bit I/O ports that may be programmed. As well, there are six counter-timers that may be used for internal and external events. A 64-character data RAM is available for logging external data upon command. Realizing these features, the I/O commands available are shown in the following table:

> LOAD PORT
> LOAD ALL PORTS
> READ PORT
> READ ALL PORTS
> READ PORT MASKED
> READ ALL MASKED
> READ PORT PREVIOUS MASK
> READ ALL PREVIOUS MASK
> AND DATA TO PORT
> AND DATA TO ALL PORTS
> OR DATA TO PORT
> OR DATA TO ALL PORTS
> XOR DATA TO PORT
> XOR DATA TO ALL PORTS
> SET BIT IN PORT
> CLEAR BIT IN PORT
> TOGGLE BIT IN PORT
> TEST BIT IN PORT
> LOAD DATA DIRECTION REGISTERS

I/O Commands

Each port is set up using the LOAD-DATA-DIRECTION-REGISTERS command. Each bit in a port may be set up to perform as an input or an output. Fig. 2-42 outlines the message structure for this command. The command code is 1E. Following this are three data characters. Each data byte represents one direction register. If a logic 1 appears in a bit position of the direction byte, that port's corresponding bit will be programmed to act as an output. A logic 0 will signify the use of an input. As you can see, each bit of the three ports is programmable using this instruction. This will only set up the ports. No data is being output to them at this time. The response to this command is also shown in the figure. It is a simple echo of the command code.

After the port is set up, data may be written to the bits that are programmed as outputs. The load-port commands will perform this. To output to a single port, follow the structure shown in Fig. 2-43. Here, depending on the port you want to write to, the command code will vary between 00 and 02. The data is the 8-bit byte you wish to show up, bit for bit, on the output pins of the part. The response message is an exact echo of the command message.

In cases where you want to send data to all the ports at the same time, the load-all-ports command fills the need. The structure, outlined in Fig. 2-44, is much like the direction register format. Three bytes of data, representing the port data, are transmitted following the command byte 03. Responding, the slave will echo the entire message.

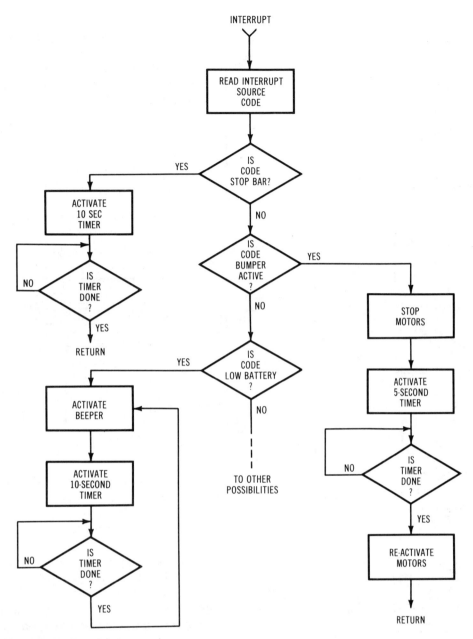

Fig. 2-35. Flowchart of interrupt structure.

Combining these two *load-port* instruction types, one can exercise a tremendous control over external circuitry at the slave site. Data and control lines may be exercised by simple commands. Adding to these, however, are even more specific port control functions. In cases where single-bit control is needed, there are three output commands dedicated to this function.

The set-bit command, outlined in Fig. 2-45, allows a single bit in any of the three ports to be set to a logic 1 value. The command byte 1F is followed by a data character that is a composite of two things. The first three bits (D0, D1, D2) will specify one of the 8 bit positions. The next 2 bits (D3, D4) will address one of the three available ports. Bits D5, D6, and D7 are unused and may be set to zeros for conve-

nience. The response to this command will only echo the command byte.

To reverse that operation, the clear-bit command (Fig. 2-46) comes in handy. Using virtually the same message format, except the command byte, it is possible to reset any bit to a logic 0. Between the two of these you would think it should be very easy to implement a condition where an external circuit needs an on-off-on stimulus. Simply sending a set-bit, clear-bit, set-bit, three-message set will suffice. There, however, Mostek Corp. has gone one better.

The toggle-bit command takes care of most of this. It will, upon execution, complement the present state of a specified output bit, thereby toggling it. The message format is illustrated in Fig. 2-47. The response is the new state of the bit.

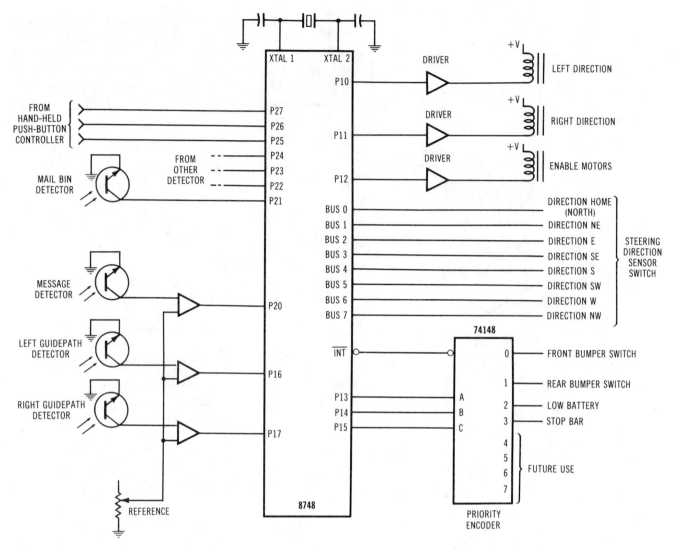

Fig. 2-36. Schematic diagram of a single-chip microprocessor-based mail-carrier robot controller.

This can then be used to toggle once again, therefore, effecting the on-off-on in only two messages.

Logical Operators

There are three logical operations that can be performed with the SCU20 on output data at each of the ports. The AND port command (Fig. 2-48) will allow data, sent from the host, to be logically ANDed with data that exists on the output pins of each specified port. The results are again latched out on that port. The response message that is returned is the new value of the port after the AND operation.

Similarly, the OR-port command (Fig. 2-49) will take data from the host and inclusive OR it with existing data from the port specified. The new value is then latched back to that port and sent as the data character in the response message. An exclusive-OR version of the same command (Fig. 2-50) also exists. To add functionality, they have added a version of these three commands that may specify the *all-ports* contingency (Fig. 2-51). In the all-ports variety, three separate data values are sent (one for each port). The response message becomes the three new values after the particular manipulations have been done.

The rest of the port commands involve the reading of data. There are simple read-single-port commands like the load counterparts, read-all commands, and a test-bit command. The latter will return the value (0 or 1) of the addressed bit in a specified port. These particular commands are all shown in the composite Fig. 2-52. They are useful for general-purpose I/O jobs and, when used in conjunction with the load commands, comprise a powerful distributed command structure. There is little that a slave unit such as this cannot be applied to in rover design. There are, however, several more commands.

Reading data through a mask is provided. This mask will allow certain bits to pass unaffected, while others will become zeros. A logic 1 in a bit position signifies a *pass through*. The mask is provided as the data portion of the read-

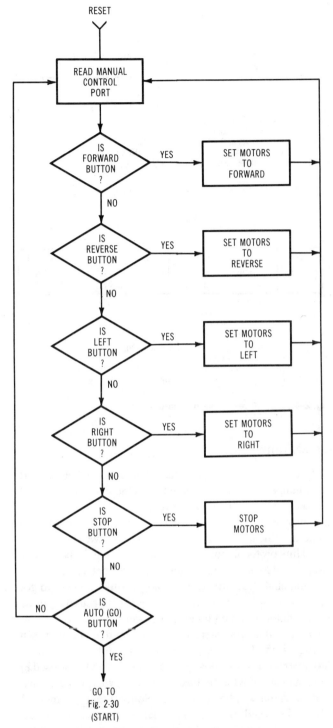

Fig. 2-37. Flowchart of initial command input loop.

handy. These commands (Fig. 2-54) send no new mask out to the ports. They rely on previously sent data. If no mask command was previously sent, all data will be passed because, on power-up, the mask register is set to all ones. If no mask command has been sent before the use-previous-mask command, these ones will still be present, and therefore, pass data.

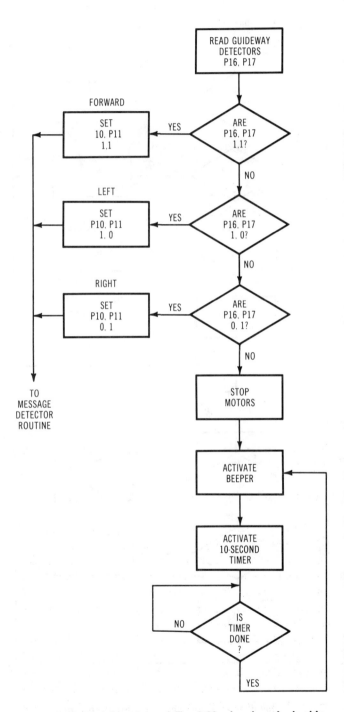

Fig. 2-38. Revised flowchart of Fig. 2-29, showing single-chip implementation.

port and read-all-ports (mask provided) commands shown in Fig. 2-53. The response from these commands will be masked data from the specified port or ports. To facilitate getting masked data from all three ports at once, the read all ports (masked) allows simultaneous operations with three masks, one per port, with only one command. In cases where the same mask may be used more than once, the read-port and read-all-ports (previous mask) commands come in

71

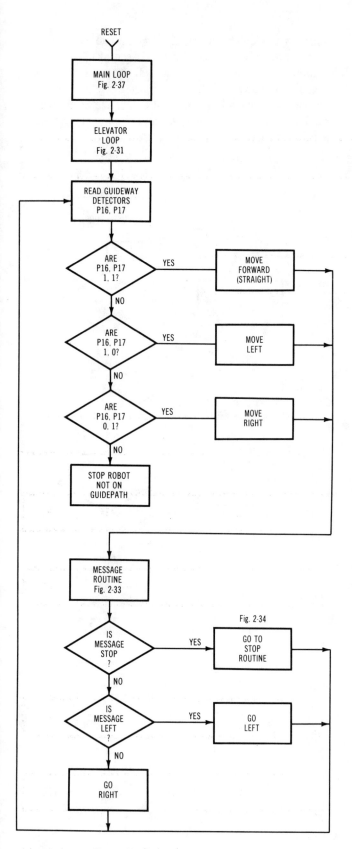

Fig. 2-39. Flowchart of single-chip mail-carrier implementation.

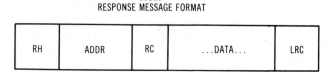

CH = COMMAND HEADER BYTE (01)
ADDR = SCU ADDRESS (00-FE)
CMD = COMMAND CODE
DATA = DATA BYTES REQUIRED BY COMMAND
LRC = LONGITUDINAL REDUNDANCY CHECK CHARACTER

Fig. 2-40. SCU20 command message format.

SCU20
RESPONSE MESSAGE FORMAT

| RH | ADDR | RC | ...DATA... | LRC |

RH = RESPONSE HEADER BYTE (02)
ADDR = ADDRESS OF SCU THAT IS RESPONDING
RC = RESPONSE CODE
DATA = DATA BYTES REQUIRED BY COMMAND
LRC = LONGITUDINAL REDUNDANCY CHECK CHARACTER

Fig. 2-41. SCU20 response message format.

Break Time

After all that, you'd think there was enough there to start designing a system. Sure there is, but there's plenty more. I wanted to take a little break to keep us on track. Realize that the distributed system we are trying to achieve must have some specification software-wise. So far, we know that a serial link protocol looks like the best way to go. The Mostek Corp. SCU20 has a well-documented error-detecting method of command structures and also appears like the way to go. From here, let's try to define some of the activities these nodes might be called upon to perform.

A theoretically advanced rover system diagram is shown in Fig. 2-55. Here, the master may be any single microprocessor system that possesses a serial port I/O to be used in communication with the link. The primary interface at each node location will be an SCU20. Notice I said "primary." There often needs to be several other circuits that, when used in conjunction with the Mostek Corp. part, will perform the function desired to meet specification. Keep in mind that the SCU20 is basically a communications link between a function to be performed and the master. It decodes messages and sends messages. The I/O port pins on the device actuate or read things external to the part.

This device is, however, capable of performing some functions at a higher level. Off-loading the master processor from tasks that are not complex yet seem to tie up processor time is the main reason for distributed systems. Many times,

| 01 | ADDR | 1E | DDR 0 | DDR 4 | DDR 5 | LRC |

Fig. 2-42. Load-data-direction-registers message structure.

| 02 | ADDR | 1E | LRC |

functions associated with the drive subsystem have shown to exhibit time-consuming bit processing. To use the SCU20 in this application for strictly bit crunching would only provide a remote I/O port. The drive system might benefit, as we have seen, from the use of a counter to, off-line, keep track of position, distance, etc. The SCU20 does incorporate counters, six of them to be exact, but to explore the commands associated with them requires us to put away the coffee and get back to work.

Timer Commands

There are five commands associated with the operation of the six counters located within the SCU20. They are the following:

Start event counter
Read event counter
Clear event counter
Stop event counter
Step event counter

Looking at them, you can see that their operation is relatively straightforward. The first action you might want to perform is to start the counter. This command (Fig. 2-56) will not only start the counter you desire, but it will set up its trigger input to respond to one of three clocking sources. There is an external input pin on the SCU20 that may be used to increment the counter. Unfortunately, there is only one of these so it may be used on one counter at a time, or, if you desire, several counters may be clocked simultaneously. As you can see in the figure, by manipulating the state of a few

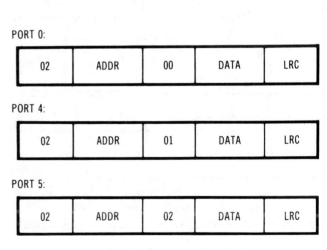

Fig. 2-43. Load-port message structure for single port.

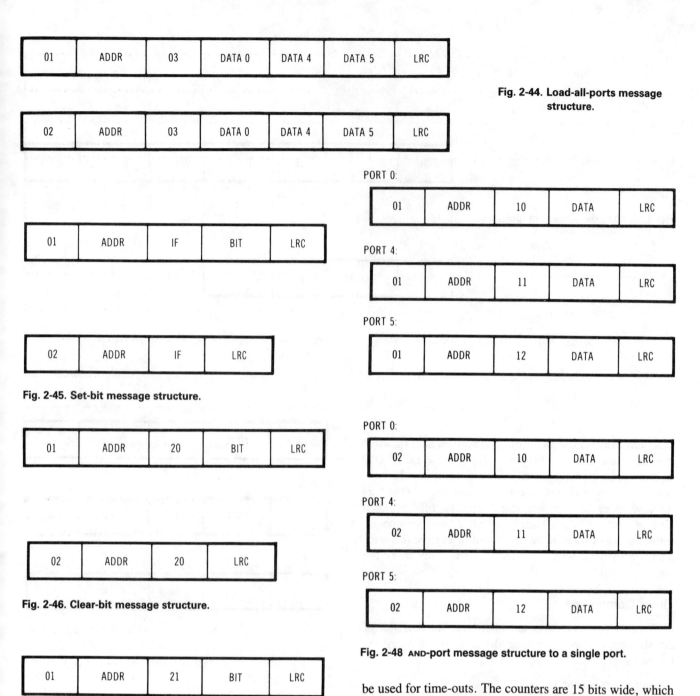

Fig. 2-44. Load-all-ports message structure.

Fig. 2-45. Set-bit message structure.

Fig. 2-46. Clear-bit message structure.

Fig. 2-47. Toggle-bit message structure.

Fig. 2-48 AND-port message structure to a single port.

bits in the command message, external increments, interval time-base ticks or step-counter commands may be used.

Internally to the SCU20, an oscillator is ticking away every 10 milliseconds. This is the internal *tick* source that may be used to increment the counters. In this way, they can be used for time-outs. The counters are 15 bits wide, which will accommodate up to 32,768 of these *ticks* before they reach their maximum value. Multiplying 32,768 times 10 milliseconds yields a counting ability spec of from 10 milliseconds (1 count) to 32,763 seconds (max count). In the case of the mailmobile, we had a requirement to count 20 seconds, beep, then count another 10 seconds. Using the SCU20 would require the master to read the counter (will be described soon) and look for 20 seconds elapsed time (read value = 4E20 hex; 20,000 dec.) then the 10 second interval (read value = 2710 hex; 10,000 dec.).

The other way of incrementing a counter is to issue a command. This command, called step event counter (Fig. 2-57) will, when received, increment the counter called for in the message. It should be noted that all timers, when they

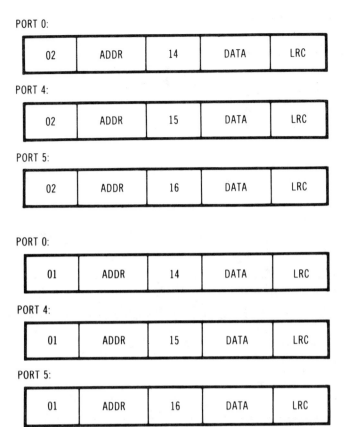

Fig. 2-49. Inclusive-OR-port message structure to a single port.

Fig. 2-50. Exclusive-OR-port message structure to a single port.

reach their maximum value, will stop. You also have the ability, through commands, to stop at any time (Fig. 2-58) or clear a counter to zero (Fig. 2-59).

Reading the timer results in a 2-byte value in the response message (Fig. 2-60). Doing so does not interrupt the operation of the counter in any way. You have to realize, however, that by the time your master software decodes the time being sent back, a new time has occurred.

Data Logging

With the timer out of the way, we are free to explore the most complex task the SCU20 can perform. As mentioned in the last chapter, it has the ability to automatically take samples of the data present at any of the three ports and store 63 of them in an on-chip RAM. The function is called its *data logging facility*.

The way it works is as follows. The start-log command is given by the master (Fig. 2-61). Embedded in it is the number of the port from which the samples are to be taken and an event counter trigger value, which will be used to determine when to actually start sampling. This value is compared against the current value in counter 1. When the two match, the data logging begins.

When logging finally begins, samples are taken at the specified port. The strobe output line (pin 7) will pulse low after the data log is started and at the end of each pass of the logging process. Each time the timer reaches the maximum count value another log begins. At any time the count of how many samples have occurred may be read using the read-log-count command (Fig. 2-62). You may also stop the log and read all samples through the use of the stop-and-read-log command (Fig. 2-63). The response message returns all values logged 1 byte after the other.

Using parts like the SCU20, and the other ones I went over, makes rover design more modular and upgradable. True distributed processing would place programmable computers at each node. Although the Mostek Corp. part is a custom-programmed single-chip computer, ones that are user programmable would affort the most versatility. Throughout this part of the software section, I have been stressing the advantages of a remote I/O scheme. There are numerous software and hardware reasons why it's better. A truly remote scheme would remote the brain from the vehicle and act on a telepathy signal much like a radio-controlled model car.

M1 SOFTWARE CONSIDERATIONS

I have constructed a relatively small radio-controlled experimental platform. The hardware description in Chapter 1 lacked any operational information. Because the vehicle has no computer, the software implications there are nonex-

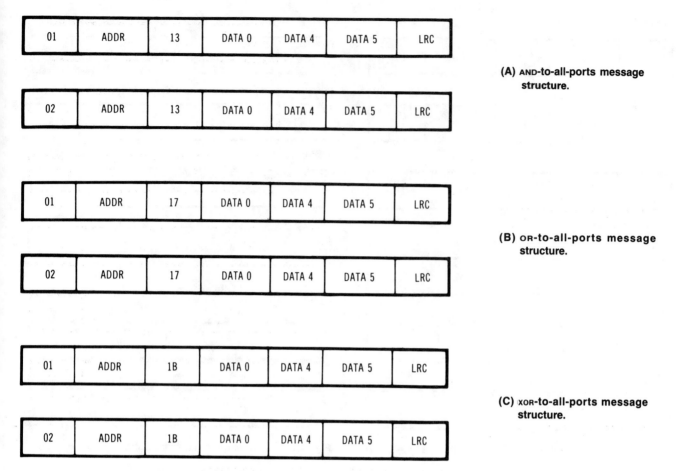

Fig. 2-51. AND, OR, and XOR to-all-ports message structures.

istent. There is, however, a larger than usual (for a rover) microcomputer equipped with a hardware interface that allows it to act as a human operator controlling the vehicle. It is the software that controls this interface that we will be concerned with here.

The first goal, while programming the interface software, was to allow a human operator, through the use of a joystick, to control the vehicle as if he were directly using the transmitter control sticks supplied with the vehicle. This may seem somewhat redundant, but it gives us a base line to experiment with the control of the vehicle through software. It will help to prove out algorithms that control the interface analog switches and give a feel for the overall responsiveness of the system.

From here, a more sophisticated set of functions may be added. A memory traceability, that memorizes all the positions the human operator moves the joystick in the order that they were moved, would be helpful to train the vehicle. This training function could include the ability to build some sort of program directly from joystick positions. This program language might consist of simple words like forward, reverse, left, right. Automatic line numbering would make it easier to read and, of course, the operator should be allowed to enter a program in the same way as now, in BASIC.

So far, the design of this software includes the following attributes:

1. Allows manual joystick control
2. Memory mode logs actions while in manual
3. Automatic program learning
4. BASIC-like program statements and controls

Along with these abilities, a feature that allows the operator a view of the current attitude of the vehicle would be nice. This view may consist of a graphic image of the rover that moves forward, reverse, left, and right in response to the codes being output to the control interface.

A program such as this has not been constructed. Let's explore the design of the program by examining each line. First, an area must be prepared to accept a program. The DIM statement accomplishes this by setting aside 100 memory areas for program lines. The number 100 is not something I calculated; it is just an arbitrary value. You may want to clear more or less than this number. After a few other initializations, a mode selection menu is displayed. It allows the operator to choose from the following:

1.) Manual Operation
2.) Auto Program Generation

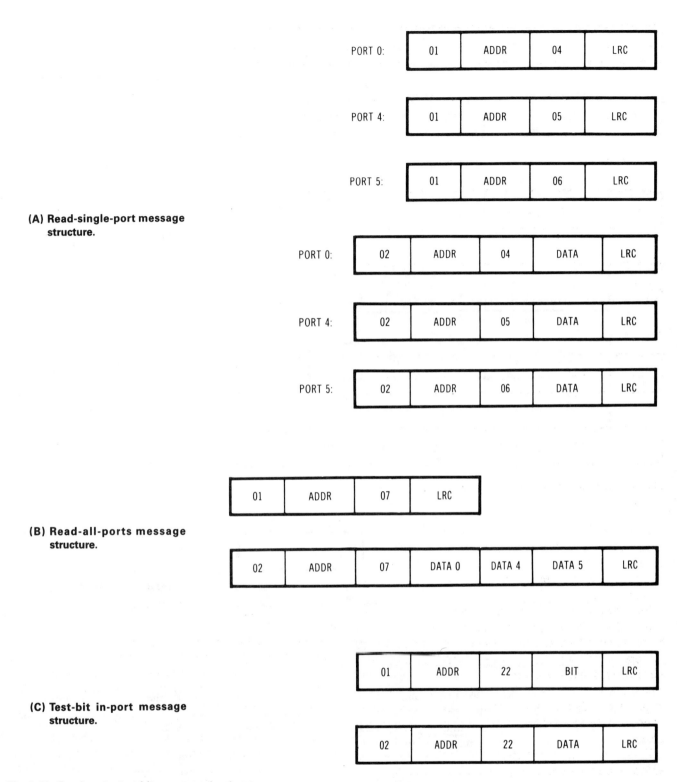

Fig. 2-52. Read-ports, test-bit message structures.

3.) Manual Program Input
4.) Load/Save Program

Each selection is treated as a subprogram. They may be designed so as to stand alone. Maybe you wouldn't want to include all options. Starting with the first, the only thing the program is to provide in this selection is the translation of computer joystick position to interface control stick value. Remember that the interface to M1 is an electronic representation of the two joysticks on the transmitter case.

The program begins by looking at the joystick ports. There are values that will be read that correlate to the stick not being deflected. As long as these values are present, the program

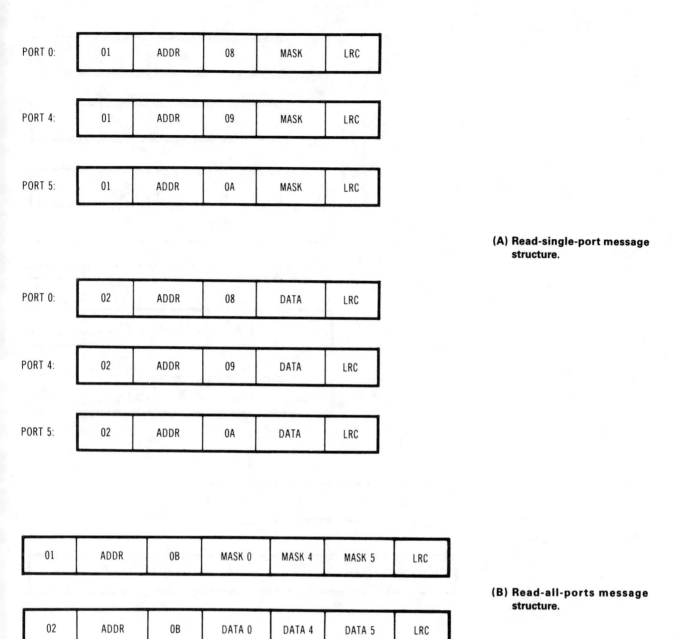

Fig. 2-53. Read-ports, mask-provided message structure.

(A) Read-single-port message structure.

(B) Read-all-ports message structure.

will do nothing. When one moves, the value is compared to a conversion table and that number is sent to the transmitter interface. It is really very simple. Typing any key on the keyboard will stop the program and return you to the main menu.

Auto program generation is a little bit more complex, but its basics come from the routines used in the manual input routine. The previously cleared memory locations for program use are numbered 1 through 100. The manual input routines are run except when a stick deflection occurs, the value is placed in the first memory location, then out to the interface. The program then marks time and increments a value. When the stick position changes, that value is stored with the first position value. The pointer is then incremented and the same thing goes on for the next program location. A keypress will stop the whole thing and return you to another menu. This one shows the following:

1.) Run Program
2.) List Program
3.) Edit Program
4.) Exit Auto Mode

The implications of all four are obvious. The first will simply reset the pointer to program location 1 and start outputting codes to the interface. After each code is output, the time-delay number is counted down until it's time to increment the pointer and get the next one. A program location with nothing in it signifies the end of the program.

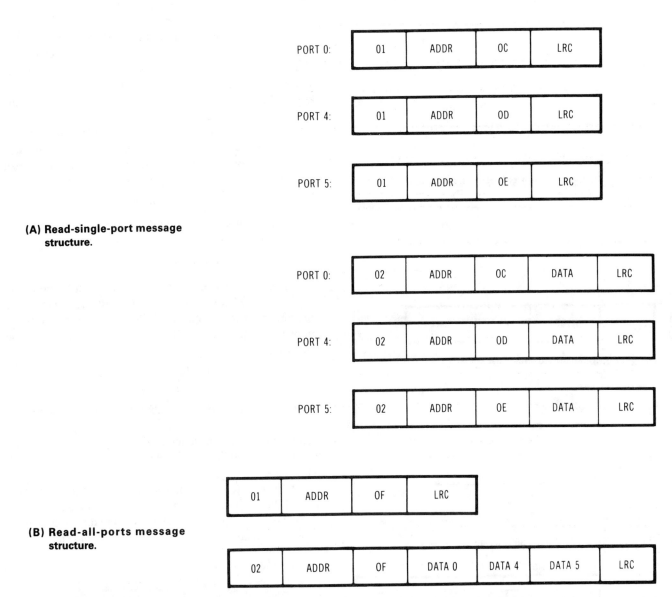

(A) Read-single-port message structure.

(B) Read-all-ports message structure.

Figs. 2-54. Read-ports, previous-mask message structure.

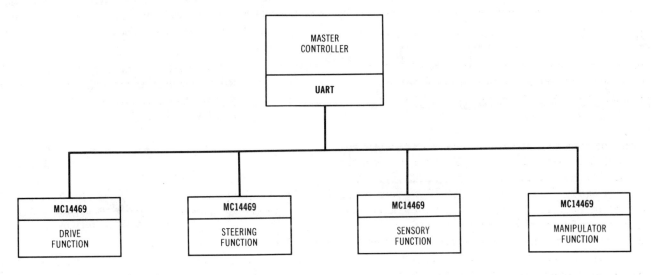

Fig. 2-55. Rover system utilizing SCU20 nodes.

79

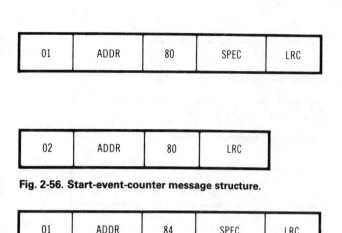

Fig. 2-56. Start-event-counter message structure.

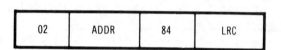

Fig. 2-57. Step-event-counter message structure.

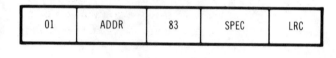

Fig. 2-58. Stop-event-counter message structure.

Fig. 2-59. Clear-event-counter message structure.

Fig. 2-60. Read-event-counter message structure.

When this end occurs, it will go back to the auto-mode menu.

Listing the program will display to either the screen or a printer the entire program existing in memory. For the screen, only 16 lines at a time are displayed. Pressing any key will advance to the next 16. The printer will print the entire listing at once.

Editing is done by specifying the line number and entering either a new value or counter number. The program will prompt you for each. When entering the mode, the screen will display:

Line Number?

After entering a valid number (you can get it by first printing out a listing), the screen comes back with:

Line xxx   RT FOR 6562

Where xxx is equal to the line number you entered. The mnemoic RT FOR stands for RighT-FORward, which is the direction the car is to go and the 6562 is representative of a number for the duration. Below the line listing will be the following:

1.) Edit Command
2.) Edit Duration

Entering one of the two numbers will allow you to change the command or duration. As you do, it is displayed on the listing line. Typing the letter Q will take you back to the auto menu. To exist to the main menu, simply select *Exit* (choice 4).

That brings us to the manual program input mode. This part of the program is very much like the editing part except there are more capabilities. The first thing the program does is ask you to enter the line number you want to start at. The screen looks like this:

Enter Start Line Number

Remember, there are only numbers from 1 to 100 available. After that, the line number is displayed on the left of the screen and the cursor awaits the input of the mnemonic direction command. These commands are composed of the following names:

```
FOR     —FORward straight
REV     —REVerse straight
RTFOR   —RighT FORward
RTREV   —RighT REVerse
```

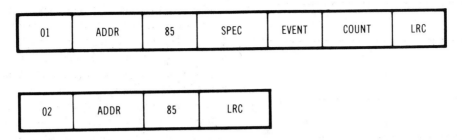

Fig. 2-61. Start-log message structure.

LFFOR —LeFt FORward
LFREV —LeFt REVerse
RT   —RighT
LF   —LeFt
STOP —No Motion

One of those nine commands will be entered as the command mnemonic. After that is entered, the duration of time the command is to be in effect is entered. This value should be derived at experimentally through the use of the *auto program generator*. Between the two, you should be able to write some fine-tuned programs. Again, typing Q at the keyboard will return you to the auto mode menu, where you have the option of running or editing the program.

What good is building a program if you can't save it somewhere? The last option allows both the saving and loading of programs. The first thing this program asks you to do is to specify a name for the program. Then you select whether to save it or load it in. When saving a program, the current program in memory will not be cleared. Loading a program, however, will clear out any existing one before bringing in the new.

Using the M1 vehicle will acquaint you with ways to direct a rover among a complex environment. It does not, however, provide for automatic control. The program just described has no provision for feedback from the vehicle. Although the vehicle has an on-board transmitter capable of signaling when a forward collision occurs, the program does not use it. Use the program as a guide in determining correct path-following algorithms. You must be the robot's eyes as you guide it. Try to establish a set mode of operation for each obstacle encountered. Later, maybe you could outfit the vehicle with a vision mechanism. Link that input device to the algorithms you have determined through manual control. Together, a great deal can be accomplished.

M2 SOFTWARE DESCRIPTION

In all aspects of the word, M2 is a robot. It contains a motion platform base with which to move. An ultrasonic vision system allows it to detect objects in its path up to 35 feet away. Above all, there is an on-board computer capable of navigating a path under program control by using these out-board devices. The processor contained in M2 is an RCA 1805. This CMOS part has more than enough capability to do the job.

I have arranged the architecture of M2 to appear as in the block diagram here in Fig. 2-64. Here, you can see that there is a port for controlling the motors (there are two) and some

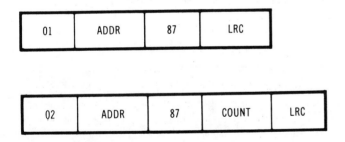

Fig. 2-62. Read-log-count message structure.

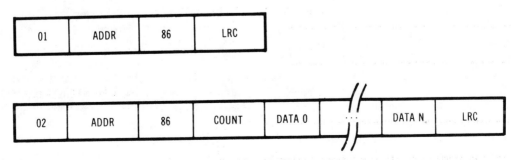

Fig. 2-63. Stop-and-read log message structure.

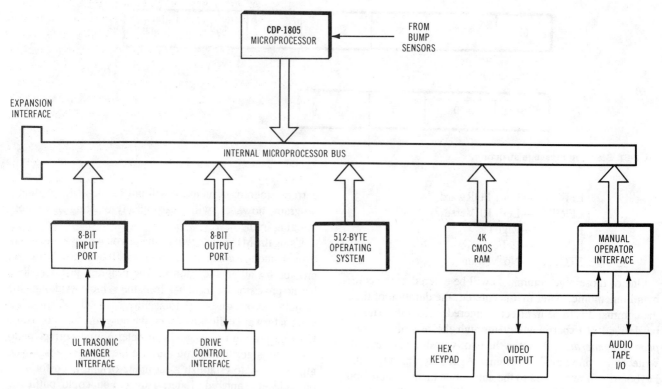

Fig. 2-64. Block diagram of M2.

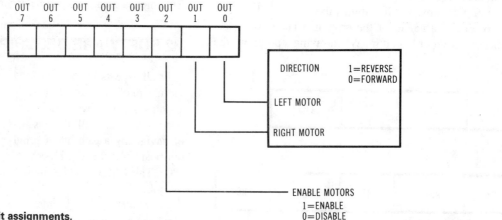

Fig. 2-65. Motor control bit assignments.

feedback as to where they are. Also shown, is the interface to the ultrasonic vision sensor. There are three digits of range, in feet, that come from that interface. The processor simply selects which digit it wants to read, then brings it in. The value read is a 4-bit BCD equivalent of the digit position value.

Let's go a little deeper. The processor is part of an RCA single-board computer. This computer contains 4K of CMOS RAM for program storage, a machine-language debugger and hex keypad for entering programs. A video monitor interface provides visual feedback of program data. Both the keypad and the monitor are mounted externally to M2. They connect only when entering programs. A cassette interface is also provided for storing and loading programs. As you can see, M2 has plenty of user-friendly computing power.

The board talks to other interfaces in the system through the use of a parallel interface. There are eight outputs and eight inputs available. The inputs are strobed into a holding register via an external latch strobe signal. There is a general-purpose output strobe called Q that is used to strobe the output information into external circuitry. An interrupt input line is connected to the front and rear bump sensors to stop the vehicle in case of an accident. Last, but not least, is an external timer input line that serves to determine the vehicle's position.

Going over the motor control system, refer to Fig. 2-65. There are 3 bits associated with its operation. One bit enables the drive system. This facilitates instant starting and stopping of the vehicle. The other 2 bits control the direction of each of the two motors. A typical sequence in programming this interface would be to first, select the motor directions and output their codes to the on-board output latch. Then, toggle the enable bit, which will start the motors. Whenever the direction is to be changed, the enable bit should be toggled, then the whole process repeated. The vehicle should pause briefly between motions to counteract any momentum buildup from the forces acting upon it during motion.

One of the motors has an integral optical dectector that straddles the drive gear. In the gear are holes that are equally spaced around the perimeter. This detector can count holes, which will relate to distance traveled back at the processor. The output pulses of the detector are fed directly into the counter input of the processor board. The timer on board the 1805 microprocessor is used to count pulses. When a command is given to travel forward so many steps, the processor will load the step count into the counter, enable motor direction and go. The counter and the detector will do the rest. When an *enable Q* instruction is executed and the motor output latch is cleared, the counter reaching zero (at position) will toggle Q, which will latch the zero motor control value out to the interface. This will automatically stop the motors.

Being able to sense objects in its path is a big advantage when maneuvering a vehicle. No longer do you have to be its eyes. The ultrasound ranging system is a modified Polaroid Ranger that has the ability to detect objects from as near as 0.9 feet to approximately 35 feet. The output of the Ranger kit is a simple three-digit LED readout. Since the 1805 microprocessor has no eyes, this readout must be converted to a digital value and signal level that can be read through the 8-bit interface.

This is accomplished easily because the chip used to display those digits takes 4-bit BCD value input from the ultrasonic control circuitry. Also, there are three digit position strobe lines that go into the display circuit. By rerouting the 4-bit BCD lines into the microprocessor input port and using one of four selectors to selectively strobe this input latch, the micro can read the digits one at a time. Two of the output port latch bits are used to select which digit to read.

Between the two, ultrasonics and motor drive, the M2 task is easy to accomplish. A program to move forward until an obstacle is within collision distance is outlined in Fig. 2-66. Following the chart, before the wheels are started, the ultrasonic ranger is checked for a collision situation. In this case, I am looking for a distance of 0.9 feet or less. If this distance is not found, the motors are commanded to go forward. After that command, the ranger is constantly checked with to detect a collision. When one happens, the vehicle is stopped.

As is the case in most ultrasonic systems, reflections and vibrations can cause false readings. Therefore, I suggest you use the check algorithm shown in Fig. 2-67 in your experi-

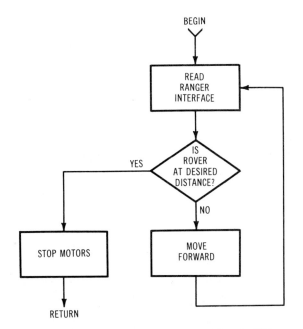

Fig. 2-66. Flowchart of simple motion algorithm for M2.

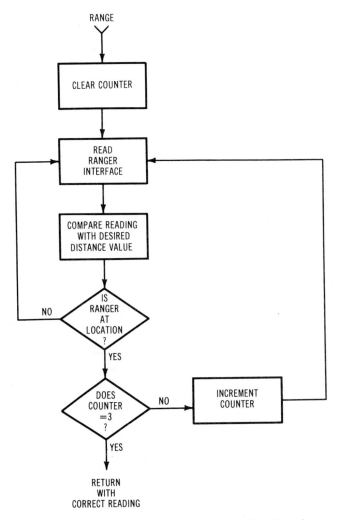

Fig. 2-67. Error defeating ultrasonic check algorithm flowchart.

ments. Here, if the value has been found successfully, it must be confirmed at least three more times before any action is initiated.

M2 is useful in proving out the algorithms you tested in using M1. You will find that, with an intelligent machine like M2, experimenting with robot vehicles becomes an easy task.

If you are truly interested in the software aspects only of the robotic profession, I would suggest purchasing one of the many personal robot vehicles now on the market. I will be describing several of these, their operation, programming, and limitations later in this volume.

ROBOT ROVERS — WHOM DO THEY SERVE?

In the beginning of Chapter 1, several present-day applications of rover robots were outlined. Throughout the last two chapters, mention has been given to all of them. The next chapter and the one following that will concentrate on *manipulators*. These are the arms that perform the work of painting, welding, and assembling in today's factories. Robot rovers are now serving as suppliers to these manipulators. The AGVs are used to ferry material to and from manipulator areas called workstations. The placement and duties of these areas is also a topic for the next chapter.

Chapter 3

Manipulator Systems Hardware

A machine that demonstrates the ability to manipulate objects around it by physically grasping them is quickly recognized as a robot. Children at the age of four can identify such machines when shown a variety of mechanical devices. Yes, manipulation seems to be a prerequisite for the classification of a *robot*.

In the previous section, I made brief reference to manipulative workstations. These areas contained one or more movable arms that were used to transfer material from the automatic guideway vehicles and to somehow transform this *raw* stock into finished assemblies. The resultant product is once again transferred to an awaiting "stock-boy" vehicle.

It is here that we begin this section. Manipulators and the systems that control their actions will be presented. To provide the most thorough understanding of the somewhat complex systems shown, I have chosen to forego in-depth discussions of mechanical design theory. Material relating to the whys and hows of manipulator construction has been presented adequately in Volume 1 of this series entitled *Microprocessor Based Robotics*, Sams catalog number 22050. Those readers desiring a deeper understanding of the mechanics involved are encouraged to obtain a copy of that volume.

THE PRODUCTION ENVIRONMENT

In a mechanized society, the effort of producing products becomes simplified. However, to arrive at a simplified method for performing this task may in itself be very complicated. In the early days of industrialization, Henry Ford found that by arranging materials so that they in effect "flow" past various workstations "producibility," or the act of producing, became more efficient. We have all heard the story of the first production line, and it won't be recounted here. However, today manufacturing engineers are once again inventing improvements on the basic *line* theory. Robots, and in particular manipulator robots, are becoming the workers at these various workstations. The production control person who used to "kit" the raw material and deliver it to these stations is fast being replaced by the AGVs described in the last section.

The manipulator systems employed in these automated factories perform a number of tasks. Some may be used as mechanics feeding raw bar or rod stock into computer-controlled milling or shaping machines. In a screw works many of the unsafe jobs once held by laborers can be dispensed by machines. These same workers can be trained to program or maintain these new "laborers." The key to this transition is in the language used to command the robots. Efforts must be undertaken to identify which methods of interaction between the former worker and the machine are the most comfortable for the human.

In the food industry, packaging of products can be handled by manipulator systems. Here sterility can be preserved and waste reduced. There is, of course, the possibility of machine grease or loose parts being introduced into the process. In these instances there exists an opportunity to train the displaced workforce to become supervisors over the machine's hands and hygiene habits. Treating the machine as the new servant and not the new master is the key to phasing in automation.

The ever-growing electronic assembly industry is now

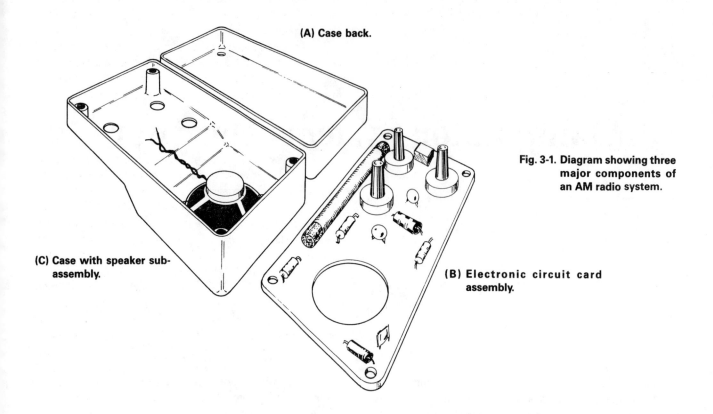

Fig. 3-1. Diagram showing three major components of an AM radio system.

(A) Case back.
(B) Electronic circuit card assembly.
(C) Case with speaker sub-assembly.

beginning to accept automation. Factories equipped with robot manipulators adept at printed-circuit-board "stuffing" are becoming more commonplace. Here, however, workers are easily trained to operate and supervise these machines because of the "high-tech" nature of the individuals involved. For these workers, automation is most likely what they produce for other industries and, therefore, they do not fear it as much as a typical foundry employee might.

In common with all these industries is the use of a manipulator robot. To the new worker of the 1980s understanding the systems used to control these robots and the methods employed to utilize them is an important step. Recalling back to section 1 where the AGV delivered material to each manipulator work station, you should begin to recognize a certain flow of products. Received raw stock, be it metal rods, bars, or sheets; powdered food stuffs such as sugar, flour, salt; or components like resistors, capacitors, and printed-circuit boards; are stored for use in the stockroom. Finished goods, be they screws, bolts, cakes, bread, home thermostats, or complete computer systems, are also stored in this stock area for shipment to other places.

The production line approach is being divided up into guideways and work cells. It is the intention of this chapter and the next chapter to acquaint you with the design and control of the work cell. Here we begin by describing some present-day manipulator cells. The design philosophy of *pick-and-place* manipulators is presented. Chapter 4 ties it all together by exploring the complete design of a manipulator language and computer simulation of the design described in this chapter.

A good place to start learning how to design a typical manipulator workstation would be to investigate the type of work performed there. Of course, it would take an enormous amount of book space to describe every conceptual duty that could be performed by a robot. Therefore, I will limit the discussion to a familiar subject: electronic parts assembly.

A product that might be representative of the field of assembly is the portable transistor radio. Most present-day radios are composed of a two-piece outer case, a single electronic circuit board on which are mounted the volume and tuning controls, and a small speaker that mounts to one of the outer-shell pieces. Fig. 3-1 is a representation of a low-cost AM radio unit. Each part may be categorized as a complete assembly.

A robot-attended workstation may be designed to perform either the final assembly or any interim task leading up to the final. In the case of the radio, the electronic circuit board will most probably be assembled by another workstation, which is dedicated just to that task. Another interim assembly might be the mounting of the speaker into one of the outer case parts.

FINAL ASSEMBLY WORKSTATION DESIGN

Fig. 3-2 attempts to depict the layout of a radio final

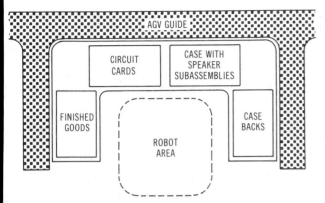

Fig. 3-2. Layout of a possible final assembly station for an AM radio system.

assembly workstation. In this view, you can see the manipulator is mounted centrally so as to be able to reach, with ease, all subassemblies. There are separate bins for case, board, and case with speaker assemblies. The robot's task might be the following:

1. Get case/speaker assembly.
2. Place on tray.
3. Get board assembly.
4. Place in case/speaker assembly.
5. Get case assembly.
6. Place on case/speaker/board assembly.

Obviously, a human doing this job would end up with a three-piece sandwich. There is no provision in the instructions to secure the board to the case/speaker assembly. Nor is there any task for the fastening of the outer shell case to the rest of the unit. To add to those problems, there is a more serious one. The speaker normally attaches electrically to the board through two wires. Does this pick-and-place manipulator need to be taught how to solder?

It is generally a rule in manufacturing industries that soldering irons are not allowed in final assembly areas. This limitation is imposed for good reason. Soldering is a slow, time-consuming task. Final assembly should be one of the speediest areas of the plant. You would expect the throughput of a final assembly station to experience a dramatic increase by introducing a pick-and-place manipulator to the work cell. If you don't allow people to solder, don't have the robot do it either.

I'm only highlighting this particular problem to illuminate the kinds of difficulties we run into when integrating robots into the workforce. In this case it's obvious that the speaker must be mounted and connected to the board prior to final assembly. In a human operation it might have been acceptable to allow the final-assembly technician to connect the wires; however, with automation, the machine may be limited in its ability to perform such diverse jobs as mounting and soldering. Much forethought must go into a product before robots can produce it.

Of course, there is a limit to how much you can change the product to accommodate automation. Somewhere along the line the robot must conform. That brings us to the design of manipulators, which is just where we should be. Will yours require grippers, magnets or maybe a suction-cup hand? Does it move around, up and down, or is it stationary? What type of control system is required for the task? Let's attempt to answer these questions by imposing another one.

Where Do We Begin?

Recall, that in Chapter 1 we were in about the same place in regard to rover systems. We began with a wish list. There we had three types of systems; here let's concentrate on the electronic assembly variety. In particular we will deal with the final assembly manipulator. You will find that the same type of control system is used throughout manipulators whether they are doing detail assembly or the rather simple task of final assembly.

Wish lists are open-ended things. You can be as daring as you want and don't have to reckon with reality for a while. In the case of our final assembly manipulator, there really isn't much we can add to a basic pick and place that would improve it. With this in mind, let's get into the basic requirements of our machine. First of all, an ability to either rotate or somehow move to at least three different places is a must. The capability of grasping the parts to be assembled is next. There are several different methods of grasping an object; however, in the requirements section just a reference to the ability is sufficient.

As you can see, that's about it. There is no need for voice I/O, ultrasonic ranging, battery chargers, etc. The robot just sits there occasionally turning from side to side to pick up an object. After assembly, the finished part is placed in an "out" bin where the closest AGV will come by and take it to finished goods. I suppose in light of this, a sort of communication should exist between the manipulator and some controller so that the AGV will come by on time, and so it will fill bins that are becoming empty.

So a wish list of demands for our typical manipulator isn't very long. Turning this list into a technical specification should be very straightforward. Because movement is the most important aspect of the manipulator's duties, particular attention must be paid to the methods used to achieve this function. In a production environment where parts are laid out in bins either around or in front of the robot, a decision must be made whether the manipulator operates in a circular fashion or in an XY cartesian mode. Let me explain further.

Fig. 3-3 is a representation of both approaches. The circular-motion robot manipulator is probably the most familiar. In this configuration the arm with gripper rotates about the vertical axis much like we would about the waist. Parts bins may be set up around the base. The arm would then turn to wherever the next part is located and lower or extend its arm. In the XY mode, all production would be performed from above and on a horizontal plane. Each part's bin would

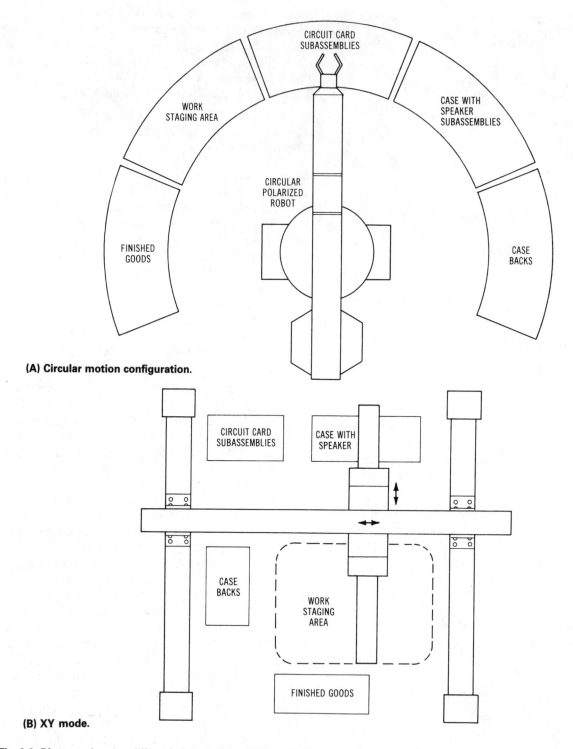

(A) Circular motion configuration.

(B) XY mode.

Fig. 3-3. Diagram showing difference between workspaces for circular and XY type robot manipulators.

have a unique location. This method is relatively easy to control except it takes up more space than the circular type on the factory floor.

Going further into the manipulative methods, consideration must be given to the reach capabilities of the arm. What is necessary? Arbitrarily, I will pick a 3-foot limit. Should it bend at an elbow or slide in and out? All the trade-offs and methods are covered in Volume 1. Let's keep the concern here on the electronic system's side. Let's assume a circular manipulator to possess the following characteristics:

1. Motor in base for side-to-side motion.
2. Motor to raise and lower arm.
3. Motor to extend arm.
4. Motor for wrist-up-and-down action.

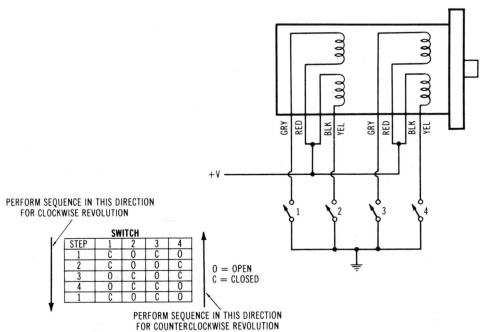

Fig. 3-4. Simple four-phase stepper motor.

5. Motor for wrist rotation.
6. Gripper mechanism (motor or solenoid).

The design of a controller board that has the capability of handling that many motors is covered in the first volume. The discussion here shall center on the incorporation of some other devices in that controller and the setup of an entire system to be controlled by software presented in the next chapter.

In fact let's get further into a discussion of the interim assembly-type robots. It is these robots that would be utilized in the making of the radio printed-circuit board. I did not cover the design of an XY manipulator in that last volume. As I said earlier, this would take up more factory floor space but requires less control theory.

Design Example

Just to cover all bases, I would like to present the design of an XY type of manipulator. I chose this type to implement because it contains many useful features, and it can serve various applications studies. Let's call it simply the *XY table* although there really is a third dimension, Z, which allows the gripper mechanism (here, an electromagnet) to be raised and lowered.

The initial purpose of this machine was to effect a test bed for doing artificial-intelligence research. A checkerboard is placed under the mechanism, and the various pieces may be moved by machine. You can imagine the possibilities. The electromagnet has the ability of *grasping* the playing pieces by attracting a small piece of metal glued to their tops. At present there is no vision system. The machine must be told the initial state of the playing field.

A little later on, I'll discuss the actual construction of the table. First, let's explore the control specification necessary to implement such a project. Let's assume that to travel in the X or Y direction, the arm needs only one motor per axis. The Z dimension requires another motor. For this design we have cut the number of motors needed almost in half. To effect simple, reliable, and accurate position control, stepper motors should probably be chosen. Electrical control of the stepper motor is relatively straightforward.

Fig. 3-4 is an electrical schematic diagram of a standard four-phase stepper motor. To move the shaft of the motor, its windings must be excited in a certain order. Connecting the motor up as shown in the diagram, and applying the control signals from the table will effect motion. A manipulator control system can do this task with software. Assuming you are using a PIA interface as described before, it is a relatively easy task to place the 4-bit control sequence on the I/O lines. Of course, some drivers must be added to provide the correct current sinking capability. A diagram of such a controller is shown in Fig. 3-5. The only possible problem of using this technique is the processor time it takes to step through and output the various codes to each stepper motor.

There are ICs now that can take the place of the software to do the step pattern and that can replace the driver parts all at the same time. This reduces the overhead on the system software. There are two parts that I have used to control the XY table. The first is the Sprague UCN4202A and the other is the Cybernetic Micro Systems CY500. Both are billed as stepper-motor controllers; however, one is much more sophisticated than the other. Let's take the simplest one first.

UCN4202A

This part is manufactured by Sprague, Inc. Basically it

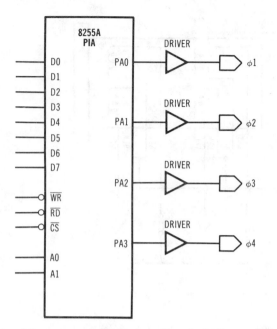

Fig. 3-5. Microprocessor interface to four-phase stepper.

performs the function of current driver and step pattern generator. Fig. 3-6 is a block diagram of the internal architecture of the chip. Its outputs can handle a 500-milliampere load at 15 volts. Most small steppers will conform to these specs. To make the motor turn, simply connect the outputs as shown in Fig. 3-7 to a stepper. Some other control source (presumably a microprocessor) has to supply a signal to specify the direction of stepping and a step pulse.

If you recall, the direction the motor will step is controlled by the pattern applied to its windings. Internal to the chip is a pattern generator that may be sequenced through forward or through reverse. Its outputs become the inputs to the on-chip current drivers that excite the motor's windings. Each pattern is sequenced one at a time through the application of a pulse at the step input. To move two steps, simply select direction and apply two step pulses. The controlling processor does not need to keep track of step patterns. In systems such as this one where there is more than one motor moving at a time, the software to keep track of the patterns can become complex.

In this case, hardware replaces software and improves system integrity. You would need some sort of driver chip anyway. This part performs both duties. Another nice feature of the UCN4202A is its ability to disable all outputs through the use of the output enable pin. Others include a step enable input to prevent unauthorized stepping and an isolated driver, which could be used to provide the current for another electromechanical device. In our case this extra driver is used to energize the electromagnetic gripper.

The part is housed in a 16-pin IC package and runs off of a standard 5-volt supply. The voltage supplied to the controlled stepper motor is also applied to the chip for the driver circuits. As you can see from the specs, the motor voltage can be anywhere up to 15 volts.

There are two modes of operation that the UCN4202A can be operated in. The monostable RC pin, when tied to +5 volts, puts the part into what is called the "full-step mode." In this mode, outputs B and D will be held stationary and each pulse of the step input will move the motor shaft one angular increment. What this means is that a motor specified to have a step angle of 7.5° will rotate its shaft 7.5° each time a step pulse is received. A double-step mode is available where the monostable RC pin is connected to an RC time constant, which effectively produces two pulses to the sequence logic each time you step once. In this mode the angular displacement of the shaft doubles (because there are two steps) and the torque characteristics of the motor increase. The full-step mode is the most common; however, double-step operation

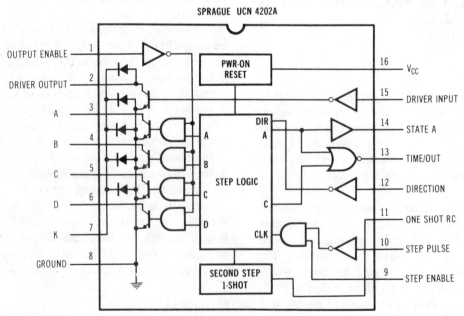

Fig. 3-6. Functional block diagram of Sprague UCN-4202A.

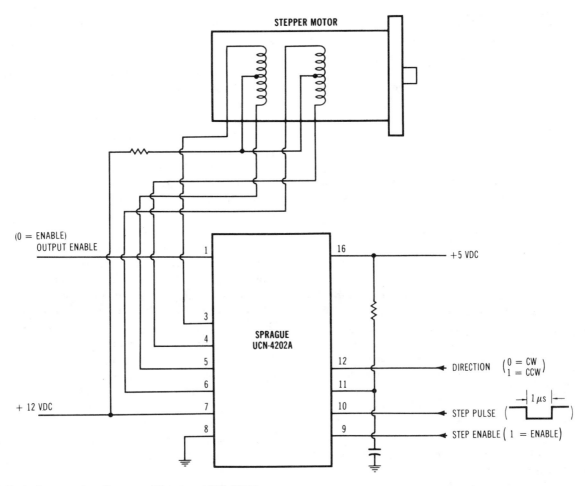

Fig. 3-7. Typical connection diagram of Sprague UCN-4202A.

may improve motor stability at high step rates.

Getting back to the XY table, there are three stepper motors. One stepper is used to pull the arm along the X axis. Another does the same job for the Y axis. Looking at Fig. 3-8 you can see the table itself. The long axis is the X direction. The shorter axis perpendicular to the X is the Y direction. The

Fig. 3-8. Photograph of experimental XY table manipulator.

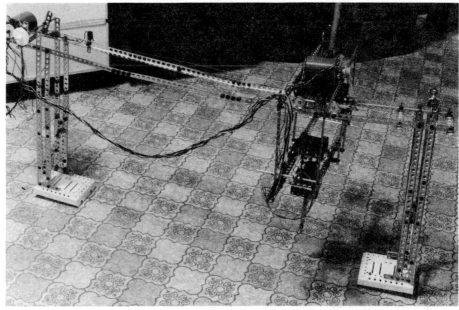

small cart that rides on the Y axis represents the Z axis and gripper. All mechanical structures are made of common Erector-set parts. The carts that move the X axis and Y axis are aluminum with grooved wheels that ride the sides of the structure. A close-up view of the Y axis cart is shown in Fig. 3-9.

Let's examine the drive structure of the X axis. Fig. 3-10 is a pictorial representation of the drive chain. A gear is mounted to the shaft of the stepper motor. This gear meshes with a larger gear that provides a multiplication of torque. The second gear drives a shaft that becomes the actual X-axis drive shaft. A small pulley is attached to the center of the shaft. Nylon fishing line is wound tightly around the pulley. One side of the line is fastened to the cart; the other side runs under it and around a similar pulley at the opposite end of the axis. The line finally ends up being tied to the other side of the drive cart.

Before the line is fastened to the back side of the cart, it is pulled taut. This tightness provides a cable drive mechanism whereby when the drive shaft is rotated, the line will either pull or push the cart down the aluminum track. The stepper motor provides precise rotational step increments that are translated into linear-motion movements. Because the motor that I chose steps in 1.5° increments, and the gear ratio from motor shaft to cart drive shaft is 3:1, the linear increment per shaft step is so small that you hardly notice it move. This prototype is capable of extremely precise positioning.

The Y axis is built up in the same way. See the photograph in Fig. 3-11. The stepper drive for the Y axis is mounted on the X-axis cart because of the weight introduced on the Y-axis cart when it is at either end. This weight (motor and cart) would bend the Y axis downward. The flexible timing belt drive works fine for transferring step rotations to the Y-axis drive shaft.

Let's stop here. It's time to talk about the Z axis. Remember that this table has the capability of picking and placing checkers around the board. In order to move around the area it is necessary to be able to lift its gripper above the playing surface. This characteristic is classified as a Z axis. To implement the linear up and down motion, I used a linear stepper motor.

If you'll recall from the first volume, a linear stepper is a stepper coil mechanism with a lead screw shaft. The front of the motor housing has a structure such that it does not allow the shaft to rotate. This shaft, however, by standing still, will cause a small bushing within the stepper coil to rotate, which effectively pushes the shaft out of the motor. The photograph in Fig. 3-12 depicts a few light-duty linear steppers. Fig. 3-13 shows the construction of the Z axis on the XY table.

The gripper, as mentioned earlier, is a simple electromagnet. It is mounted on the shaft of the linear stepper Z axis. The magnet may be energized or de-energized to pick and place pieces. Of course, the magnet must be maneuvered into place over the piece first.

Talk about maneuvering brings us back to control. There are three stepper motors and one electromagnet in the system. Another addition is a warning flasher, which would mount atop the X-axis cart, and a warning beeper mounted within the control box. These two devices indicate when the axes of the table are about to move under program control. They serve as a safety mechanism to warn people around the unit.

From the requirements set forth so far, it is obvious that three UCN4202A parts will be necessary to control the table. Fig. 3-14 is a complete schematic of the control interface. Each part controls one axis. The isolated driver outputs of these parts control the gripper, warning flasher and buzzer. Digital control inputs to the interface come from a computer system. There are three separate direction and step controls and one common output enable line. A simple parallel interface could provide these signals from a variety of computers.

Three outputs go back to the control computer from the table. These represent the states of the limit switch of each axis. When an output is low, the indication means that the corresponding axis limit switch has been activated. These limit detectors are mounted at one end of the track of each of the X and Y axes and at the full up position of the gripper. They are provided to tell the computer when each axis is at its home position. By the way, *home* in this system is the upper right corner of the checkerboard. This location is referred to also as 1,1. Where the first digit represents the X axis and second the Y axis. You will see, in the next chapter, where you will be able to command the table to move its gripper anywhere around the checkerboard by simply calling out the X,Y coordinate pair.

Remember, I mentioned at the start of this section, that there are two devices suited to controlling this table? Well, the second is much more sophisticated than the first. In fact, you don't give this one any step pulses or direction control line signals. A simple ASCII command message will send the table seeking a specified XY location. Let's explore the workings of the circuit.

Cybernetic Micro Systems CY500

The CY500 is designed to control a stepper motor from a master controller through the use of either immediate commands being sent to it or from an internally stored sequence of instructions. All commands come into the CY500 as ASCII characters, therefore, making it easy to interface to. In fact, for prototyping an application, the CY500 may be connected to a standard ASCII keyboard. Its parallel data input will allow TTL signals to command the processor. Did I say processor? Yes, the part is actually a custom-programmed 8048 microprocessor. When you connect the CY500 to a computer output you have the ability to control the motor's actions as if it were *on-line* to the master.

There are two modes of operation of the part. The command mode is likened to the immediate mode of most programming languages. In this mode commands received will be carried out immediately. A programming mode is

Fig. 3-9. Photograph of X-axis cart.

included where each command received is stored in an on-chip buffer for later execution. Because of its ability to enter the programming mode, the CY500 is called a *stored-program stepper-motor controller*.

To my knowledge, no other component exists that can perform stepper motor control from an internally stored program. This greatly reduces the system software tasks relating to manipulator control. The ability to send a string of commands to the part, command it to start execution and then go off and perform other duties is a tremendous help. Let's look into the internal structure of the part.

Inside the chip there are six functional subsystems.

1. Input data system.
2. Output data system.
3. Program parameter storage.
4. Mode flags and user I/O.
5. Program storage buffer.
6. Instruction selection, decoding, and control.

Commands coming from the host master enter the input data system. These commands may be entered via parallel or serial data. Fig. 3-15 shows a typical parallel hookup to the CY500. In the figure you can see that a 7-bit data path is connected to the parallel input pins on the port. A write strobe input will, when pulsed low, signal to the part that data is available on its input pins. A low, true, busy flag output indicates to the communicating master the current status of the device. Using a standard D-type flip-flop, a simple parallel output keyboard may be attached to the input, which facilitates prototype program development.

As I said, a serial input option is also available. A communications mode pin (PAR/SER pin 39), when tied low, will program the device to accept serial ASCII commands using the write strobe input (pin 1) as the command line. Once again, the busy input should be consulted before transmission. This can be accomplished by connecting the busy

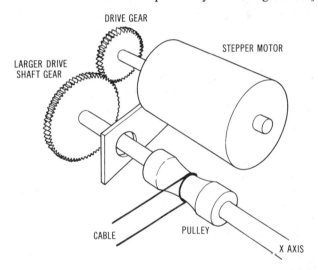

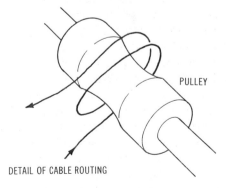

Fig. 3-10. Diagram of cable details for X-axis drive.

Fig. 3-11. Photograph of Y-axis assembly.

output to the RTS input of the communication UART. A drawing depicting a serial hookup to an RS232 host is included as Fig. 3-16.

Now that we know how to hook up and command it, how do we connect the motor? Well, the output data system does include four controllable outputs for just that purpose. They conform to the sequence code chart shown previously for stepping the motor. External current driver devices (shown in Fig. 3-17) must be supplied. This particular device is made by Sprague and is very similar to the one incorporated in the UCN4202A. Notice also that an inversion stage is provided between the CY500 and the driver.

There is a separate programmable output pin (pin 34) that may be used for any purpose the user desires. In a manipulator system, this output could be used to control an external parts handler or conveyor. Simple uses could include the activation of a warning flasher or buzzer. As we saw in the construction of the XY table, both of these devices are used. Another application may be to perform the activation of the electromagnetic gripper.

Since we now have commands entering this part and a stepper motor and electromagnetic gripper connected to it,

Fig. 3-12. Photograph of linear stepper motors.

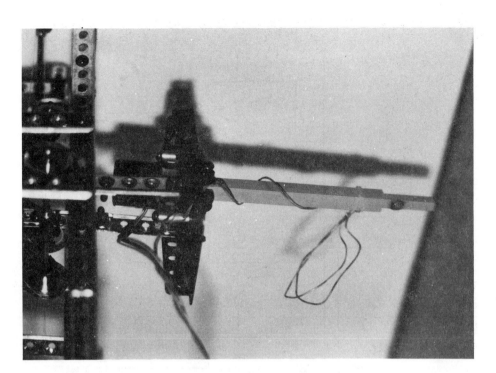

Fig. 3-13. Photograph of Z-axis assembly utilizing linear stepper motor.

let's see where everything takes place inside that 40-pin package. Fig. 3-18 depicts a logical block diagram of the CY500. To show an actual diagram would simply duplicate the internals of a single-chip microcomputer. Since the control program of this part has been customized to where it acts as a dedicated function, it makes sense to view it as such.

This point in discussion brings us to the program-parameter storage section. In this section, control entities such as step rate parameters, step-ramp parameters, and data pertaining to the current position register are stored. You see, controlling a stepping motor can be much more complex than simply sending coil energization patterns to the windings. Each step may be executed at a certain rate. There is a maximum allowable speed at which each motor may be stepped. The step rate parameter will configure the part so that it will execute each step according to a set rate.

As it is with many electromechanical devices, the physical properties of the stepper motor beg to be started slow and to be slowly *ramped up* to the desired step rate. This approach will ensure a more consistent torque at the load-bearing end. The step ramp rate parameter allows the programmer to select the operational ramp that the motor is to follow. The CY500 is smart enough to perform a ramp function according to this parameter.

While stepping, the CY500 will update a 16-bit counter, which keeps track of its current absolute position. This counter may be used in some instructions to compare with a desired absolute position received from the host. A second 16-bit counter is used to accumulate relative position data. This counter will increment or decrement an amount equal to the number of steps taken since the last stop point. Relative comparisons may also be performed when an instruction specifying them is received.

The mode flag section consists of various inputs and outputs designed to control the operation of and current status of the CY500. One of these control inputs has been previously discussed. The PAR/SER input configured the CY500 for either type of communications input. Other inputs include a reset signal that will initialize the part. During initialization any internally stored program is erased, motor direction is set to clockwise, motor outputs are de-energized, and the command mode is entered.

Data may be entered in either ASCII character format or in straight binary. A step rate value of two steps per second would be specified as a 32 in ASCII format. This value corresponds to the equivalent of the number two. The binary representative of the value would be 00000010, where the logic 1 in the second bit position indicates a binary weight of the value two. Communication in either of the two modes is selected via the ASCII/BIN mode control input, pin 33.

Whenever a catastrophic chain of events occurs, due either to mechanical failure or incorrect programming, it is possible to stop the CY500 from executing that fatal next step. The abort input line (pin 6) will, when low, immediately stop all actions on the outputs and return the part to the command mode, looking for the next instruction.

So far, we have been able to talk to the device, command it (through electrical means) to initialize, and stop at a moment's notice. We have also been able to select which way the data enters the chip (parallel or serial) and the method of communications (binary or ASCII). There are other inputs that allow a greater degree of control. These controls, however, may or may not be active depending on the overall mode of the device. Let's go over their actions assuming they are enabled.

Fig. 3-19 shows a view of the CY500 as it is configured to

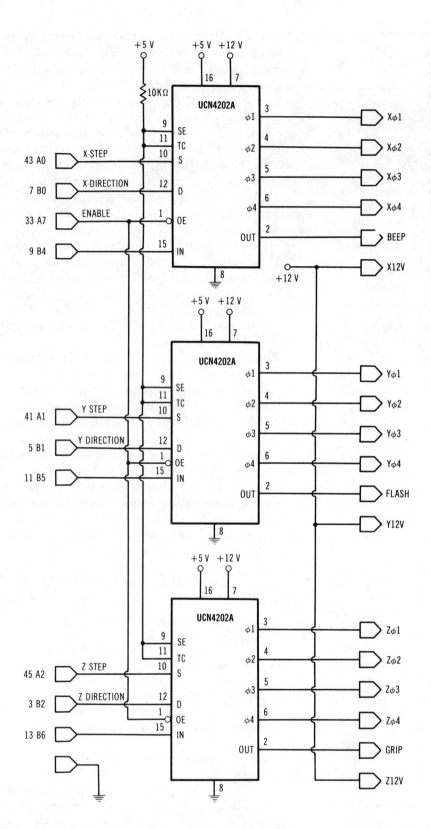

Fig. 3-14. Schematic of general-purpose XY table interface using UCN-4202A stepper-motor controllers.

date. Notice that there are a few new input lines present. The trigger input (pin 30) controls when the next step is to occur. This pin may be likened to the step pulse you sent to the UCN4202A. When low, the device will step along. If the input is high, the part will simply await a return to the low level.

External direction (pin 29) will signal to the part the direction of motor movement. This command may be performed over the communication port; however, in some modes the state of this pin determines direction. External

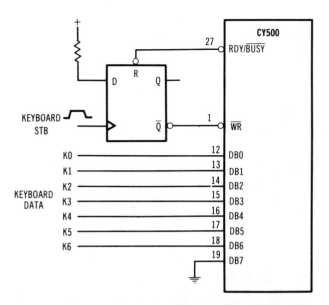

Fig. 3-15. Diagram showing connection between Cybernetic Micro Systems CY500 and a keyboard for input.

start/stop is another function that may be initiated by software. As you can see, the part is set up so that it may be operated manually.

There is only one more command input line left to be discussed. The wait input is a general-purpose signal line that may be connected to a variety of things. There are commands that the CY500 may be configured by that will perform or not perform duties awaiting the state of this line to change. In the case of finding the home position of the X, Y, or Z mechanisms of the XY table, the wait line may be connected to one of these limit switches. A set of commands may be entered that will send the motor stepping backward toward the switch as long as the switch is not closed. As soon as it is closed, a command is issued that tells the position counter that this is the home position (ATHOME).

To go along with these mode-control inputs, there are a few status outputs that are helpful in determining the state of the part at any given time. Two of these outputs indicate the stepping motion. No, they're not the same as the four-coil outputs. The toggle and pulse outputs will change state on each movement of the motor. The toggle output will change state (up or down) each step. The pulse output produces a position 5-25 microsecond pulse each step of the way. These can be used to synchronize other external circuitry.

Whenever the CY500 is in the program mode, the program mode output pin 31 will go low indicating the device is ready to accept stored data. On the other hand, when the CY500 is in the control mode, this output will switch to a logic one corresponding to the new mode of operation. When the device is running a program, the RUN signal transition to low will indicate this fact.

The one remaining output can be used as an interrupt to the main system. The motion complete (pin 37) signal will indicate that all motion from the previous command has finished. This line can be used as a handshaking input to a master controller. By polling this output, the master can determine when a given command has terminated.

This description of the CY500 is by no means complete. As you can imagine, there is a great deal of software command support that is involved with the part. We will pick up in the next chapter where we leave off here. You now have an idea as to what the inputs and outputs are. Fig. 3-20 is a complete electrical pinout of the part and the following figure

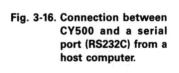

Fig. 3-16. Connection between CY500 and a serial port (RS232C) from a host computer.

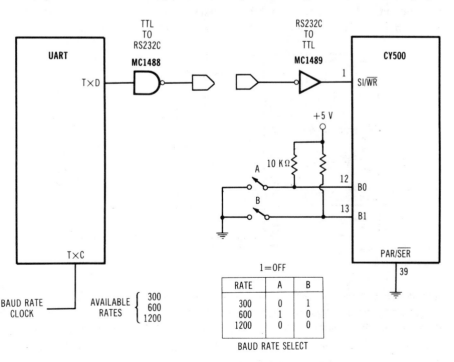

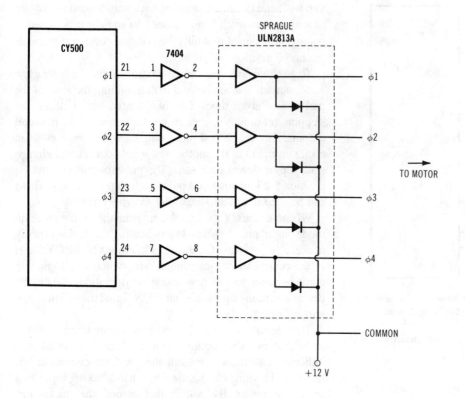

Fig. 3-17. Interface between CY500 and a stepper motor.

(Fig. 3-21) shows three of them connected together to implement the XY table electronics.

Before we got deep into the use and description of the two stepper controller parts, we were generalizing on the use of manipulators. The XY table is indicative of what is called a *pick-and-place* robot. This type of mechanism would not be utilized in the final-assembly robot we discussed initially. The radio circuit board may be assembled, however, by this type of robot. To take it one step further, the stockroom where all these parts emerge and are sent back to will use a vertical adaptation of the XY table to retrieve pallets from storage bins.

The next chapter covers a great deal. The command structure of the CY500 will be explored and completed. Then, a complete program will be designed utilizing the UCN4202A to control the table. Input for this program comes either from an immediate operator (keyboard) or from another computer system. This other system will be designed to simulate the

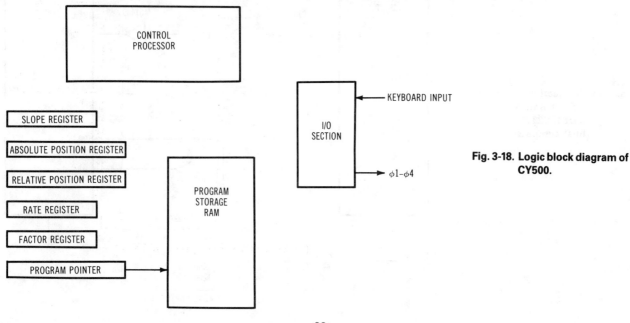

Fig. 3-18. Logic block diagram of CY500.

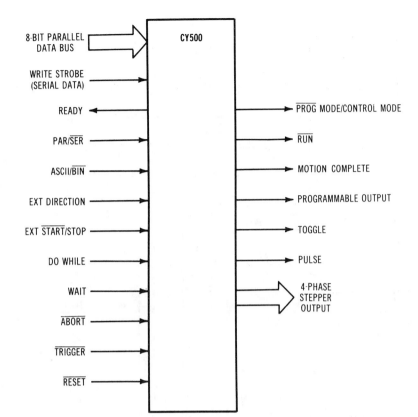

Fig. 3-19. Diagram showing remaining input control lines.

mechanical actions of the table through the use of computer graphics. All programs will be described as we close out on the subject of manipulators.

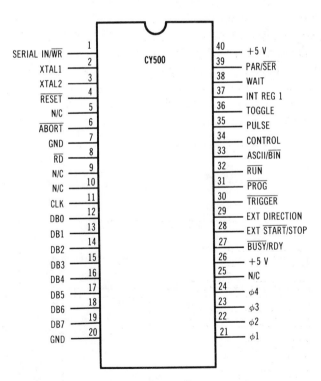

Fig. 3-20. Pin designations of the CY500.

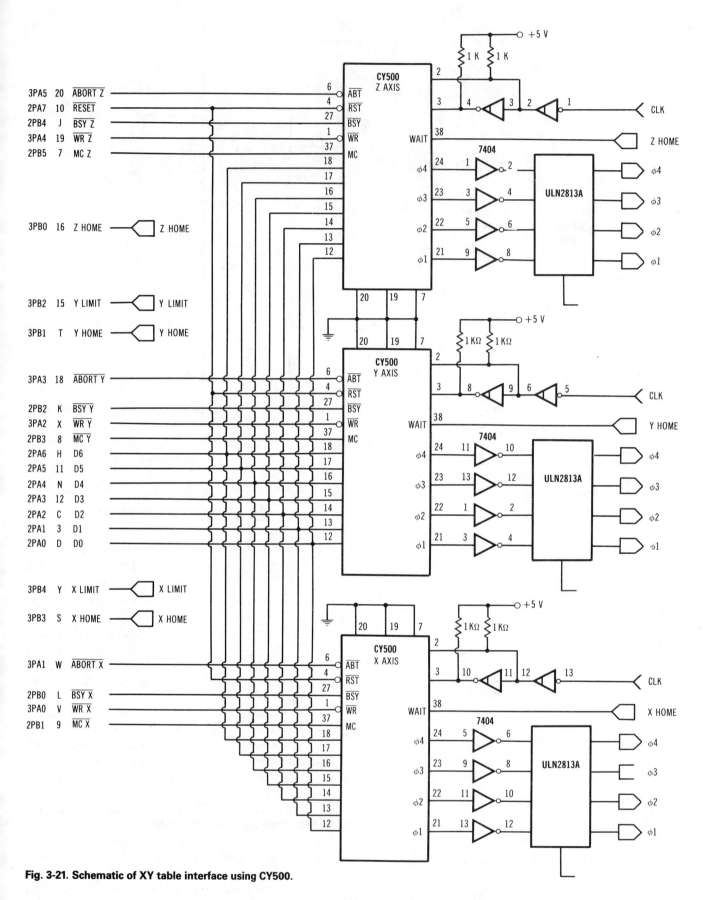

Fig. 3-21. Schematic of XY table interface using CY500.

Chapter 4

Software for Manipulators

Between the last chapter and several of the chapters in Volume 1 of this series, manipulator control by hardware and software means has been well covered. The purpose of this chapter is to close out the discussion of the single-chip stepper controller (CY500) from a software viewpoint and to introduce the concept of manipulator programming languages.

Recalling our discussion in the beginning of Chapter 3, we posed several examples of the utilization of manipulators in industrial environments. The final assembly station was conceived by citing an example using the manufacture of a small portable radio. As the discussion went on, our thoughts turned to an XY table more along the lines of a pick-and-place robot. In all the examples, the prime objective of the manipulator was to perform physical actions that, when put together, accomplished a specific task.

Regardless of the mechanics used to perform manipulation, some human command link had to be established somewhere along the line. Some manipulators are self-contained and, therefore, require no external computer attachment. Others are merely slave mechanical subsystems. The XY table is a prime example of the latter. When a computer controller is involved, a command language must be established to convert the task from human terms into parameters that the robot can understand.

This command language must travel somehow into the manipulator subsystem. There are several established paths that are open to utilize two different electrical interfaces. One of the interfaces is the standard RS232C serial communications link. Over this, data is sent as a series of bits that are represented by two different voltage levels. The other interface employs simultaneously two handshaking signals. Whichever you choose doesn't really impact the design of the language itself. Let's explore the design of a manipulator language.

COMMAND LANGUAGE DESIGN

If we treat the design of this language like the design of any part of a robot system, the first task to perform is to provide some sort of wish list. Before we do that let's consider the overall goal of the language. If the purpose of the design is to communicate a task to the manipulator, then the specific method must be defined. There are several levels of communication. The highest may come across as simply "assemble the radio." Although this may sound simple and to the point to you, there is a great deal of information conveyed in those three words.

Splitting the sentence up into three parts, let's examine the first word. "To assemble" is to put together. There are probably thousands of ways to perform an assembly task depending on the parts involved. Giving a command as vague as "assemble" assumes that the manipulator has, at its demand, a vast library of assembly methods, task descriptions that are sufficiently coded so as to present the mechanics in the system with an appropriate control command. Depending on the library, this may be an impossible task. It certainly is not the place to start dabbling in language design.

The next two words, "the radio," specify the piece parts that the assembly operation is performed on. This part of the

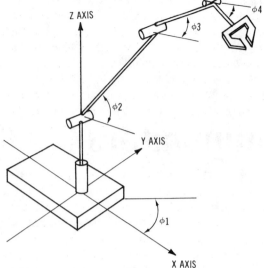

Fig. 4-1. Circular coordinate system.

sentence serves to further define the tasks necessary to assemble. The word "the" is probably unnecessary and may be discarded. All it is used for is to make the human feel comfortable when specifying a command in English. As is the case with the first word, the word "radio" may be too vague a description. If the manipulator were to have a library of radio descriptions, the command interpreter would be tremendously complex.

Adding a further complication, where is the robot to find "the radio" on which to assemble. Does this imply some sort of vision capability on the robot's part? Perhaps the radio pieces have been loaded into precisely defined holders by another machine. The latter would seem the likely choice; however, that simply moves the sophistication of locating and placing the radio pieces off to some other machine.

Obviously, a lower-level command structure would be the place to start designing command languages. That's not to say that after you master the lower levels, a high-level interpreter cannot be attempted. After all, it will be necessary to define a lower level even in that high-level interpreter. There are several other levels of command. The very basic level would employ the actual step commands to the motors. Depending on the structure of the manipulator, this level may be indistinguishable from another similar-level structure.

We have covered the final assembly robot, which basically has a work area where piece parts are laid out for the manipulator to grasp and "assemble." This type of robot may work from a single base where the manipulator may have the freedom of turning from side to side to reach the work place. A robot such as this works on a circular coordinate system. Fig. 4-1 depicts this type of system. As you can see from the figure, a point in space is defined as a series of angular positions of each joint. For a manipulator to be flexible enough to perform the task of radio assembly, the amount of joints in the arm turn out to be between five and six. To specify each angle for six joints becomes rather complex.

Assembling the PC board that goes into the radio would be carried out by another type of manipulator. The pick-and-place XY approach requires no joint angle positions. Each coordinate is specified as simple X,Y cartesian points. To simplify the system even further, any negative points may be eliminated by specifying an outside corner of the manipulator's reach as location 1,1. From there all other positions will be positive. Fig. 4-2 shows the relative coordinate plane of the XY table type of manipulator.

Because an experimental mechanism exists to prove out XY commands, this is the type of command language we will

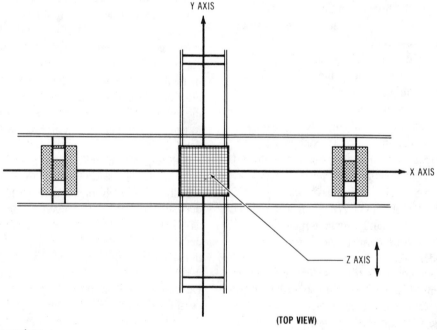

Fig. 4-2. XY coordinate system.

deal with primarily. Much of the design, however, is adaptable to all manipulator systems. In those areas where extensive modifications would be necessary, I will further explain the differences.

Specific Commands

The first place to start, as mentioned earlier, is with a wish list of the functions you would like the language to perform. In preparing this list keep in mind the limitations of the mechanical subsystem you are using. Also, think about methods that can improve operation, accuracy, and simplicity. Let's get into the following, which is a typical wish list of the XY table:

**XY Table
Command Language
Wish List**

- Local keyboard operation
- Remote operation from host computer
- Automatic program entry mode
- Ability to exit program

The first item we encounter is the ability to operate locally. What this means is that you can sit at the keyboard and command the table in real time. What type of commands would be necessary? Well, let's list a few. The following would be absolutely necessary.

1. MOVE X,Y
2. HOME

The arguments X and Y in the move command would specify the place you would like the arm to go to. The HOME statement allows you, in one command, to move the arm back to its place of origin. In this system that place would be classified as location 1,1. Between these two commands you have the power to move that arm anywhere within a particular work space. Given the fact that the arm has the power to retrieve and place things, you might want to add two more commands.

3. PLACE X,Y
4. GET X,Y

With these two commands, you now have the ability to grasp objects such as the radio pieces or pieces that go into the radio printed-circuit board and place them into other areas of the work space.

Now that we have the MOVE, HOME, PLACE, and GET commands, a small command language can be designed. How do you get these into the system? Realizing that we're still working in the local mode, we have to work out a system for entering key commands through the keyboard. Obviously, the easiest way of doing this would be to use the English words *move*, *home*, *get*, and *place*. The X,Y arguments would then be specified as decimal numbers. It is here that we have to start defining, a little further, the work space. In the XY table example, a standard checkerboard, 8 × 8 or 64 distinct work-place squares are used. You could use any type of system that works well for you. Going along with the 8 × 8 mechanism, the X and Y values would allow one number from a range of numbers from 1 to 8 to be placed in them. Therefore, a typical command would be MOVE 1, 8. That would tell the arm to move to X location 1, Y location 8. A GET command, specifying GET 4,4, would physically move the arm to location X-4, Y-4 and the Z direction would be moved down appropriately to pick up the block or whatever object and bring it back up to its highest location. The PLACE 1,6 would then take that block from the previous GET command and place it at X location 1, Y location 6.

As you can see, by adding commands a complete sequence of events or perhaps even an assembly method could be designed. The next item on the wish list is to have a remote operation capability. This would allow another computer system to interface with the arm controller mechanism. Using similar commands, this external computer system could control the arm just as if you were doing it from the keyboard. It would be wise to utilize the same type of command structure as was used in the local mode. This cuts down on the programming of the entire application.

Programming codes, specific to this type of mode, will be different depending on the interface communications path that you have chosen. In the example that will be shown later, a serial RS232 command path is connected from one computer into the computer that is controlling the arm XY table. As mentioned earlier, a parallel path would be just as efficient.

Let's move along to the third item on the wish list. Program mode will allow the operator to enter in a sequence of MOVE, HOME, GET, and PLACE commands from the keyboard. These commands, however, will not be carried out until the program is run.

Entering these commands in the program mode is very much like entering commands in standard BASIC. You would want the computer to ask, when you enter this mode, whether a new program is being entered or an old one is being edited. When a new program is being entered, an automatic line-numbering system should be designed so that the user does not have to enter in numbers before commands. After the MOVE X,Y command is typed and the enter button is pushed, a new line number should come up.

When designing such a system, an end delimiter of some type must be provided for. In this case, I have chosen to use the letter Q. When Q is entered, the program mode is exited. After entering a program, you must have the ability to either edit it, list it, or run it. You might even want to be able to store the program off on some off-line storage device such as a cassette or disk.

All these command possibilities will be shown later in the text. The last item on the wish list is, obviously, to be able to exit the program at any time. This could be done just by entering in a number on a menu.

Let's talk a little bit about controls in the possibility of an emergency situation. Even though the manipulator XY table that is being built is relatively low power, small, and table top, and does not involve the possibility of endangering human life, those types of mechanisms must be thought of at this stage in the game.

A manipulator system must have the ability to signal that it is moving. We touched on this in the last chapter with the beeper and movement flasher. There is another mechanism that, in the command language phase, must be brought into play. This is that of the limit switch. It is possible for an accidental command to send the arm past its mechanical limit. This may not, at this stage of the game, result in any injury to humans but it might become a catastrophe for the mechanical device. Therefore, limit switches at all end points on the X, Y, and Z axes must be provided for. There should be the possibility of aborting any command at any time. When in the local, remote, or run program mode this mechanism must be provided for.

Another area to discuss now is that of *coordinate conversion*. As I mentioned, in the English command system that we have outlined, the X and Y arguments are allowed to take on values of 1 to 8 in the decimal system. The stepper motor or the motor controller that drives the stepper motor of the XY table does not necessarily understand the decimal value 1 to 8 to mean an actual place in the work space. Therefore, some type of coordinate conversion must occur.

Assuming that you are using the UCN4202A stepper-motor controller, the coordinate conversion becomes a little more complex. To move from location 1 to location 2 on the 8×8 block set may require several steps of the stepper motor. It may even be as high as 32 steps depending on the size of the blocks.

A conversion table must be built in software that converts the decimal number to the number of steps, in which direction. Also, when you are at location 4 and 4, and the next command makes you go to 3 and 4, the coordinate converter must be intelligent enough to realize that there is a change in direction on the X axis, which results in only one set of steps instead of three sets of steps. As you can imagine, this can get very complex. An algorithm to convert the decimal numbers to steps will be shown as we go through the actual design program.

Well, that about covers the wish list. Of course, you might have added other items. This wish list is just the basic outside entities. It is not a specification. An actual specification for what we have just described may look like the following one. Here, some of the little items, including the running of programs, the number of commands, etc., are listed.

Specification for
XY Table Command Language

1. Operation Modes
   1a. Local
   1b. Remote
   1c. Program
2. Commands
   2a. English
   2b. Moves
   2c. Gets
   2d. Places
   2e. Program Commands
3. Interfaces
   3a. Drive for three stepper motors
   3b. Keyboard interface
   3c. RS232C host interface

From here, let's look into the two different methods of stepper motor control. In the hardware section of Chapter 3, we discussed the design and application of the Sprague UCN4202A stepper motor controller and the electrical characteristics of the Cybernetic Micro Systems CY500 stored program stepper motor controller. As you may remember, there are advantages to using each. The easiest system to use is the Sprague unit; however, the computer that is running the device must hold all the burden in software. The CY500, on the other hand, requires little software. A simple coordinate or statement converter could be designed that downloads new commands into that intelligent chip.

Because of its simplicity, the first unit that we will go over is the UCN4202.

Sprague UCN4202A Initialization

When we begin to address, or command, the XY table, the first thing that we must do is to assure that the arm is at its home position (X location 1, Y location 1). There are a few ways of doing this. As I mentioned in the last chapter, one method is to step the X motor in the opposite direction while constantly checking for the 1 position limit switch to be closed. When this limit switch is activated, stepping stops, and the next MOVE direction is changed. This is very simple to do with the UCN4202A. A routine that sets the reverse direction, which will depend on the way you have connected up your stepper motor, and that sets the step pulses stepping, will accomplish this easily.

When the X direction has reached its limit, the same type of routine is started in the Y direction and then the Z direction. When you are sure that all limit switches are set and activated, that the arm is at location 1,1 and the Z direction, and that the pick-up coil or the electromagnetic on the end of the XY table gripper is at its upmost position, most of the initialization has been completed. Now you would want to set the outputs that control the directions to the opposite direction so that all you have to do from there on is prepare for the next step. This constitutes the setup portion of the control routine.

Through experimentation, you should be able to find how

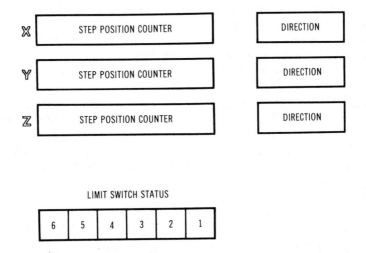

Fig. 4-3. Programmer's model of a typical stepper interface.

many steps it will take to go from one work-space square to the next. This will depend on the gear ratio of the stepper system and the number of degrees per step of the motor that you have chosen. In my particular system there are 32 steps per block. Each block is approximately 1 inch square. I have set the limit switch so that the arm, during initialization, stops at the center of square 1,1.

A world model could be set to where each space has a certain number assigned to it. For instance, space 1,1 would have the number 0, space 2,1 or X location 2, Y location 1, would have the number 32, the next 64, and so on. Therefore, when we tell the arm to move in the X direction 4 spaces, it will know to go to location 128. This type of a system would be called a *world model*. Every location would have two numbers. It's no different than the one through eight world model that will be built into the command language program except that this system has actual steps associated with each block. The control system running the arm is embedded in the control system running the language so, basically, you have a subprogram that has its own world model.

Also, you must employ, what I would call *step counters*. These counters will allow you to step the amount of distance that you require. For instance, if you want to go to location 128, you would set the counter to the value 128 and then just step. Each step would then decrement this counter. When the counter got to 0, you would be at location 128.

Communication between the UCN4202A subprogram and the main program should be through coordinates and direction. The UCN4202A subprogram should only be employed to physically change the state of the direction, and step and enable lines of the three stepper-motor control chips on the interface board. No other high-level command structure should be employed in this subprogram.

A programmer's model of a subprogram such as this is shown in Fig. 4-3. Here we have three step counters that can be used to position the arm, and we have three direction registers. Notice also, that there are six limit switch locations that, depending on their state, signify which end limit the arm is presently at. The actual design of the UCN4202A program will be shown later when we discuss the full command-language design. As you can see, there really isn't much to say about the Sprague unit.

Cybernetic Micro Systems CY500

I promised, in the last chapter, that the short discussion of the Cybernetic's CY500 would be expanded upon in this chapter. Because it is an intelligent IC, most of its actions are carried out by software commands. Therefore, at this time, let's get into the CY500, go over each command and find out exactly what this part can do. In the last chapter we learned how to hook it up to the three stepper motors in the XY table and how to communicate with it. The actual command structure is what we'll deal with now.

Much like the command language structure, the CY500 allows you to enter ASCII commands from a keyboard or from a serial data link, which will be converted into step commands to move an arm. In this aspect, the CY500 could be called a high-level language device. There are two ways to enter commands to the CY500. One is a single ASCII character followed by a carriage return. The other is a single ASCII character followed by a blank, then an ASCII decimal number parameter, and then a carriage return.

Some examples of the types of command are shown in Table 4-1. Let's look at that table now. The first command we come to is ATHOME. The command is signified by, simply, an ASCII A with a carriage return. What this does is tell the part that at the present time its location is 0. It is performing the initialization step in software that we went through for the UCN4202A. No, it does not move the arm, it simply tells the part that it is at the home position. We'll come back to this.

Table 4-1. Examples of the Types of Commands to the CY500

SINGLE-BYTE COMMANDS		
NAME	COMMAND	DESCRIPTION
ATHOME	A <CR>	DECLARE ABSOLUTE ZERO LOCATION
BITSET	B <CR>	SET PROGRAMMABLE OUTPUT LINE
CLEARBIT	C <CR>	CLEAR PROGRAMMABLE OUTPUT LINE
DOITNOW	D <CR>	DO PROGRAM
ENTER	E <CR>	ENTER PROGRAM MODE

TWO-BYTE COMMANDS		
NAME	COMMAND	DESCRIPTION
NUMBER	N n <CR>	DECLARE NUMBER OF STEPS TO BE TAKEN
RATE	R r <CR>	DECLARE MAXIMUM RATE PARAMETER
FACTOR	F f <CR>	DECLARE RATE DIVISION FACTOR
SLOPE	S s <CR>	DECLARE RAMP RATE
POSITION	P p <CR>	DECLARE TARGET POSITION

The next command is BITSET. As you may recall, there is a single programmable output line available on the chip, which is not used for any specific function. This output line may be used to turn on the beeper or the flasher, or even be used to cause the arm to grip. The BITSET command allows you to turn on this output. Conversely, the CLEARBIT command, which is the next one on the table, specified by an ASCII C and a carriage return, allows you to clear that output line. Through the use of the BITSET and CLEARBIT, you now have the ability to grip or ungrip a block.

The next command in the table is called DOIT. It should be obvious from the wording of this command that this will instruct the chip to do or begin running the program that has been stored. By now you're asking, what program? If you recall in the last chapter, we said that there is the possibility of storing a bunch of these commands, 18 in particular, into an on-board memory. Performing this DOIT, or ASCII D and carriage return, command will then run that program.

The next command, ENTER (ASCII E carriage return), will put you into the program mode. While in this mode you can store single-letter or double-letter commands. Speaking of double-letter commands, let's move down on the table to where there are five more examples (Table 4-1B), keeping in mind that this is not the full command set. The first example that we come across is called NUMBER. The ASCII command is N, space and the actual decimal number, in ASCII values, of steps to be taken. Now, these steps are relative to where the motor is at presently. This is not an absolute command such as, "Move the motor to step 128." This would be, for example, "Move the motor 9 more steps or 6 more steps or whatever."

The RATE command, or ASCII R, specifies the speed at which the stepper will step. We talked about this briefly in the last chapter. We will go over it in more detail as we cover the complete command list. The next one, FACTOR or ASCII F, sets a division factor of the rate. Both the rate factor and the next one called SLOPE all interrelate and will be explained later. The last command on this sublist is called POSITION. This command will declare the absolute mode of operation. If you recall, the NUMBER command will give you the relative amount of steps to step. The POSITION command, on the other hand, gives you an absolute position, just like the 128 was in the UCN4202A command structure.

That's only ten of the commands that are available on the CY500. At this time, let's go into the full list. Table 4-2 is a complete list of all the commands in alphabetical order. At this time, you'll notice some redundancy as we have already gone over ten of the commands. We'll go through them again for clarity.

First one on the list is ATHOME, specified by an ASCII A and carriage return. The ATHOME instruction defines what we are calling the home position. In the XY table, home is 1,1. Home here could be the 1,1 position; however, within the chip this position is called absolute 0 and is a reference for all position commands.

There are two modes of possible failure when using this command in the current CY500. One is that you should not use a change of direction command immediately before the ATHOME command. The second is that you should not use ATHOME command twice during operation unless the part is initialized or reset prior to the second ATHOME command. This is why ATHOME will typically be used during the initialization phase when the power is turned on and the arm is set to its home position. The only other time that HOME should be used would be to actually step the arm back to the home position. The ATHOME location should not have to be specified more than once.

The next one, BITSET (one-byte instruction, ASCII B,

Table 4-2. CY500 Command Summary

ASCII CODE	NAME	DESCRIPTION
A	ATHOME	SET CURRENT LOCATION EQUAL ABSOLUTE ZERO
B	BITSET	TURN ON PROGRAMMABLE OUTPUT LINE
C	CLEARBIT	TURN OFF PROGRAMMABLE OUTPUT LINE
D	DOITNOW	BEGIN PROGRAM EXECUTION
E	ENTER	ENTER PROGRAM MODE
F	FACTOR	DECLARE RATE DIVISOR FACTOR
G	GOSTEP	BEGIN STEPPING OPERATION
H	HALFSTEP	SET HALF-STEP MODE OF OPERATION
I	INITIALIZE	INITIALIZE SYSTEM
J	JOG	SET EXT. START/STOP CONTROL MODE
L	LEFTRIGHT	SET EXT. DIRECTION CONTROL MODE
N	NUMBER	DECLARE RELATIVE NUMBER OF STEPS TO BE TAKEN
O	ONESTEP	TAKE ONE STEP IMMEDIATELY
P	POSITION	DECLARE ABSOLUTE TARGET POSITION
Q	QUIT	QUIT PROGRAMMING/ENTER COMMAND MODE
R	RATE	SET RATE PARAMETER
S	SLOPE	SET RAMP RATE FOR SLEW MODE OPERATION
T	LOOP TILL	LOOP TILL EXT. START/STOP LINE GOES LOW
U	UNTIL	PROGRAM WAITS UNTIL SIGNAL LINE GOES LOW
+	CW	SET CLOCKWISE DIRECTION
–	CCW	SET COUNTERCLOCKWISE DIRECTION
0	COMMAND MODE	EXIT PROGRAM MODE/ENTER COMMAND MODE

carriage return) is, as we said before, the command that makes the output pin (pin 34) go high. It is a general-purpose output. We may use it as a flasher, beeper, or gripper command. Keep in mind that these are MOS-level, TTL-compatible signals and cannot be used to directly drive motors or electromagnets.

CLEARBIT (one byte, ASCII C, carriage return) will reset the output pin 34 to a logic low condition. The ASCII D carriage return DOITNOW or DOIT is as we specified before. The instruction will cause the CY500 to begin executing the program that would have been previously stored. If no program has been stored, the controller will jump right back into the command mode. Basically, it will ignore this command.

The interesting part about this command is the fact that it can be used within a program. For instance, if you have a program that you wish to continuously loop, the DOITNOW would be placed within the stored program at the end of a loop of commands. Whenever the DOITNOW is encountered, it will immediately restart the program. Therefore, you could get continuous loops.

The next command is ENTER. It is a single-byte command specified by an ASCII E and carriage return. This will cause the CY500 to enter into the program mode. From then on, all commands that are communicated to the part will be stored inside the program buffer.

Let's skip over the FACTOR command and go right on to the GO command. GO will cause the stepper motor to step after you've already specified the rate, direction, and things like that. Rate, direction, number of steps, and such parameters are typically entered in real time in the command mode.

They are usually not a stored entity. After doing something like POSITION, which we will go over again, you are specifying an absolute value for the motor to step to. You would have already specified the rate and the factor that rate will be divided by. The GO command is the actual doer, the command that makes the chip perform the steps.

The next command, HALFSTEP, is a one-byte command specified by an ASCII H, carriage return. The HALFSTEP command will put the CY500 into a mode that resembles the single-step command of the UCN4202A. Each step is very small, as specified by the motor angle parameter. If the motor is specified as a 1.5° stepper, each step will be 1.5°. A smoother step rate and action will happen during the HALF-STEP mode of operation.

The next command is called INITIALIZE. The INITIALIZE command will bring the CY500 back into the command mode if it had been in the program mode. INITIALIZE will basically reset the part and, as I said earlier, would be necessary if the ATHOME command is to be used twice. What should be mentioned is that none of the distance or rate parameters that have been stored prior to the INITIALIZE command will be altered. So any commands after the INITIALIZE command that have not respecified distance or rate parameters will be performed at the prior parameter values.

The next command, although single byte, JOG, enters the part into a totally different mode. It becomes a manually operated stepper controller. When the JOG command is started, external start/stop control is initiated. Remember the pin external start/stop. When the J command has been entered, from then on the application of a logic 1 at the external start/stop pin will cause the connected motor to begin stepping, and it will continue stepping until that pin is brought low. In this way you can manually move the position of the motor. Before this command is entered, however, valid rate and directions commands must be entered so that the part knows the speed at which you are to step. Position is not necessary for the external JOG command.

The next command is LEFT/RIGHT pin enable or ASCII L, carriage return. This particular instruction is very much like the JOG command. It now places the CY500 in the external direction control mode. In this mode the direction of stepping is controlled by the application of a high or a low on the external direction pin (pin 29). As you will see later, there are commands that can specify direction. When you are in this mode these commands are ignored.

Next command is NUMBER. Now, we've already gone over the NUMBER command a little bit. Let's go over it in a little more detail. The NUMBER command is used to specify the number of steps to be taken when you are in the relative mode. It is a three-byte command, ASCII N, space, and then the least significant byte and the most significant byte. The argument for relative number of steps can be anywhere from 1 to 65,536, which is basically a 16-bit number that is 65,000 steps from the present position. If you are to store this in the program buffer, realize that it takes up 3 bytes out of the 18-byte buffer.

The next command is called ONESTEP. It's a single-byte command, ASCII O, carriage return. This command will take one single step. It will work at the rate that has already been specified when the ONESTEP command was sent. This allows you to fine tune the position of the arm.

Speaking of position, that's the next command, ASCII P with two following bytes. POSITION command is, as I said, the absolute position you wish the motor to go to. Each time the stepper motor is moved, the CY500 keeps track of the current position. Even when a relative command is entered, the current position register within the part is updated. If, for instance, the relative position command has stepped you to location 127 and now the POSITION command asks you to move to 127 when the GO command is entered, because the absolute position requested is equal to the current position no step will be taken. By the way, the argument following POSITION may be as large as the one following the relative numbers command, or basically 65,536 possible locations in the work space.

The next command is called QUIT. It is a single byte. It basically tells the part to stop the programming mode of operation. You will see the Q used in the command language that we design for the same purpose. It will cause the CY500 to exit the programming mode.

We're going to skip a couple of instructions here. We're going to skip RATE and SLOPE and move on to the LOOP-TIL or single-byte T instruction. This is a high-level type command that may be likened to the programming command DOWHILE. It has the capability of running a program until a

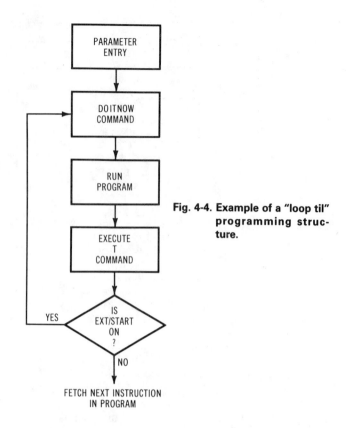

Fig. 4-4. Example of a "loop til" programming structure.

certain condition on an external pin is met. I mentioned this briefly in the last chapter. The EXT input, pin 28, if low, will cause this command to loop back to the beginning of a program. You would use the T command in the program mode. Fig. 4-4 is an example of a program flowchart written with the T command. As you can see, from the figure, DOITNOW will start running the program. When the program encounters the T command it will check the state of pin 28. If pin 28 is low, it will go back and execute the DOIT-NOW instruction, which will start the program again. If the pin is high, it will then go to the next instruction in the program, which might be a QUIT instruction. This is the type of command that you would use when initializing and moving to location 1,1. Let's explore that a little more deeply.

Let's say that the 1,1 limit switch or the X-1 limit switch has been connected to pin 28. The command computer sends down a series of instructions that tell the CY500 to go in the reverse direction 65,000 steps relative and then the T instruction is put in, which will, basically, say as long as the limit switch is low, or normally closed, keep on stepping. When the switch is high or activated normally closed to open, the next command will be the ATHOME command. This is signified in Fig. 4-5. Using this program flowchart, the unit will automatically seek its home positions if there are three CY500s on one board and they are all connected up to their respective limit switches, sending the same program down to all three and then executing the commands will automatically initialize the XY table.

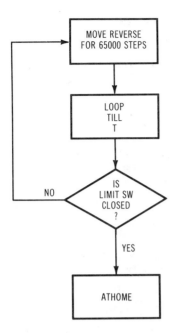

Fig. 4-5. Example of an "auto-home" routine.

The next command, WAITUNTIL, or the ASCII U, is very similar except that it uses pin 38, which is the WAIT input. Either the T or the U command could be used for initialization. Of course, there are other uses for the U or T commands depending on what you are trying to sense.

The next command is the PLUS command or clockwise command. All steps following this command will be taken in a clockwise direction. Conversely, the MINUS or counterclockwise command will change direction.

The last instruction in the list is the ZERO command. The ZERO command will place you in the immediate mode or command mode of operation. In this mode commands are executed immediately and are not stored in the program buffer.

Now it's time to go back to those commands that we skipped over. The reason behind the skip-over is because the RATE, FACTOR, AND SLOPE commands can become relatively complex and should be treated separately. Steppermotor control systems can be designed very simply or very complexly. The more complex they are, the more exact they are. The CY500 was designed with the possibility of implementing both structures. The rate control portion of the CY500 allows very precise positioning of the motor. You can control the step rate over a range varying from one step every 5 seconds to over 3,350 per second. It is, however, nonlinear and the resolution increases at the slower speeds.

Now let's get into the RATE command. As you can see from the table of instructions, the RATE command is an ASCII R followed by a space and then a single-byte parameter. This parameter is in the range of 1 to 253. It corresponds to rates of 49 steps per second to 3,360 steps per second. Now, there are two formulas that allow you to determine this parameter from the actual steps per second that you want to achieve. One formula works for parameters in the range of 0 to 200:

$$\frac{12500}{(256 - R)}$$

The other formula works for parameters greater than 200:

$$\frac{12500}{(257.344 - R)}$$

Table 4-3 allows you to pick the nearest value of parameter without using the formula.

Table 4-3. Rate Table

PARAMETER	FORMULA	EMPERICAL
0	48.8	48.8
25	54.1	54
50	60.7	60
75	69.0	69
100	80.1	80
125	95.4	95
150	117.9	117
175	154.3	154
200	217.9	217
205	238.8	239
210	264.0	264
215	295.2	296
220	334.7	335
225	386.5	387
230	457.1	457
235	559.4	560
240	720.7	721
245	1012.6	1013
250	1702.0	1702
251	1970.4	1970
252	2339.0	2339
253	2877.0	3359

It gets more complicated. Speed range, discussed thus far, assumes a division factor of 1. It is possible, however, to divide the basic rate that has been specified up to 255. This allows you to get less than 49 steps per second. As you can imagine, stepping at a rate of 49 steps per second may be too fast for most applications and may be too slow for some applications. The ability to go less than 49 steps per second is done through the FACTOR command. As mentioned, FACTOR and its parameter argument may divide the rate up to 255. A table showing typical rates and the factor parameters employed to achieve them is shown in Table 4-4.

Now that you have the rate control specified and the factor of the rate, the only command left to talk about is SLOPE. SLOPE is also called *slew rate*. Because the CY500 is capable of very high slew rates, it may be impossible for the electromagnetic motor to achieve that rate when it is initially stepped. The SLOPE command will instruct the CY500 to variably increase the speed from a known start point. The slope parameter tells the part to increase the rate by the slope parameter with each step until the maximum stepping rate is

Table 4-4. Rate Parameter

RATE DIVISOR FACTOR	STEPS/SEC	SEC/STEP
1	48.8	
2	24.4	
3	16.2	
4	12.2	
5	9.7	
10	4.8	
20	2.4	
24	2.0	
48	1.0	
99	0.50	2.00
100	0.50	2.05
150	0.33	3.07
200	0.25	4.09
245	0.20	4.99
250	0.20	5.10

Coordinate Controller Command Set

Program Commands
- Edit line
- List Program
- Run program
- Program mode

Movement Commands
- Move arm
- Get block
- Place block
- Home
- Goto

Setup Commands
- Plot current world
- Plot target world

reached. This rate parameter defines the jump between rates. Between these three instructions, it is possible to have complete, precise positioning of the motor by ramping-up in speed and then ramping-down.

That about does it for the instructions for the CY500. As you can imagine, putting these commands together can provide you with a tremendous number of possibilities for XY table or any other manipulator system control. But let's get into the driver mechanism that would be used to control the CY500 from our command language.

The initialization I've gone over would consist of that series of commands that, if connected up to the WAIT or TIL input line, would automatically initialize the XY table. The world model could be handled in a conversion subprogram that lists, much like the UCN4202A world model, actual absolute values for each work-space square. The communication link between both subprograms could be the same. In fact, in the beginning of your command-language handler, you may have a prompt that asks you which type of controller you are talking to.

It's now time to move into the actual design of a command language. The initial language we are going to go over is implemented in MicroSoft BASIC. The specific machine was a Radio Shack TRS-80® Color Computer. The XY table is utilizing 4202A stepper-motor controllers and CY500 stepper-motor controllers. Prompts are shown on a video screen. The local mode is handled via the keyboard of the Color Computer. The remote mode is talking through an RS232 serial communications port to a Texas Instruments Computer system that will be described later.

Let's get into the program. Before we go deep into the program listing, let's begin by defining a few more commands. I've taken the liberty to expand on the GET, PLACE, HOME, and MOVE for reasons of simplicity and to allow a greater degree of control over the manipulator. In functional order, the new commands are listed in the following:

Let's go over them one by one. The EDIT command works basically the way an EDIT command would work in a BASIC or higher-level language. It allows you to go into a preprogrammed or an earlier programmed line and edit the command that is on that line. In this particular case, you will replace the whole command on that line with a newly entered command entered from the keyboard. Because each one of these commands are so short it's not worth designing an actual editor down to the character level.

Another command, EXECUTE, we are going to leave for now; I will explain it in more detail later. GOTO is exactly the same as in a BASIC language. As I said before, there will be line numbers on the stored program that you are allowed to enter from the keyboard. The GOTO will allow you to, unconditionally, jump to one of those line numbers.

GET is the same as was described before. HOME is also the same. LIST will list out the program that you have stored in the command buffer. It will list out and ask you if you want it to go on the screen or the printer. If you decide to send it out to the printer, it will ask you the name of the program so that it may label the printed sample.

MOVE and PLACE are as we discussed earlier. PLOT is a new entity, and I will describe it later when I go over EXECUTE. PROGRAM will put you into the program mode, which was explained a little bit earlier. RUN will run the program stored in the program buffer. TARGET goes along with the PLOT and EXECUTE commands. That's it for the list. Now it's time to come up with EXECUTE.

We talked some about world models. This particular command example or command language is at a higher level than the actual language that steps the motors. It is possible to put a little higher-level command structure into the world model. The world model for this program is a series of black-and-red pieces just like checkerboard pieces. It is possible for us to build a world model of an 8 × 8 workspace array in which black or red entities may be stored. The PLOT program will allow me to tell the computer, workspace section, what is located in the blocks. Because I am assuming you have no vision interface or any means for the arm to see what the block looks like, it will be necessary for you to initially, when you first run the program, tell the computer what the state of the world is at the present time. The PLOT command asks you, square by square, what is in each workspace.

The TARGET command will allow you to set up another world model. This world model would be used as a to-get-to place. For instance, the state of the world may be one way, you would like it to look a different way. Through the use of the EXECUTE command the arm would automatically be able to change the current world into the target world. In the listing provided, EXECUTE is not programmed. There is a reason for this, and I will explain it at the time we get to the EXECUTE command. So that's the list of commands for the Color Computer, what I call the *coordinate converter*. It takes instructions from you, the human, or from another computer system like the Texas Instruments system and commands the positioning of the XY table.

Now let's get into the screen displays of the program. In this day and age of commands placed on a video screen, the command placement of a video screen is very important. The coordinate converter program allows you to select from a menu, in the beginning, and then go into a basic screen with a simple prompt that allows you to enter in the commands in English. The basic screen will look the way it is shown in Fig. 4-6. On initialization you have four possible modes of

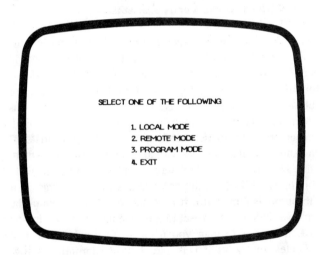

Fig. 4-6. Representation of the initial program screen prompts.

operation. If you'll notice, these four modes look very much like our wish list. The local mode will send you to a routine that takes input from the keyboard. Remote will connect you to the Texas Instruments' system. Program mode will send you into a program storage routine that we will go over soon. The last mode allows you to exit from the program. Exiting from this program, seeing that it is written in BASIC, will send you back to the operating system of your computer.

The initialization phase of the program is shown in Fig. 4-7. I'm going to explain, line number by line number, the

```
10 CLS
15 DIM CW$(8,8):DIM TW$(8,8)
20 S$=".":E=1:DIM PL$(51)
25 POKE 49153,9:POKE 49155,9
30 POKE 49152,255:POKE 49154,255
35 POKE 49153,13:POKE 49155,13
40 POKE 49152,255:POKE 49154,255
```

Fig. 4-7. Listing of initialization routine.

actions that the program is performing. After the title remarks of line numbers 5, 6, and 7, line number 8 is a POKE to a memory location. If your BASIC does not have a POKE instruction, some sort of direct writing to a memory location statement would be used. The POKE statement uses the decimal value 65,407 as an address, 0 being the data to be sent to that address. This is machine specific and selects a parallel I/O interface. My number 10 clears the screen for operator interface. Line 15 begins the initialization of certain variables.

The DIMension statement will clear away a certain area of memory for later use. The CW string corresponds to the current world model that will be used. Notice that it is eight locations wide by eight locations deep. Similarly, a dimensioning of a target world or TW string also has 64 locations arranged 8 × 8.

Next line (line 20) describes S string as being an ASCII period. You will see this being used as a delimiter in the command sequence. E is an error indicator, being 1 is always showing an invalid command. You'll see this in use later also. And then a final DIMension statement for program line strings or PL string and we allow 51 program lines cleared away.

The next four lines are machine specific. What they are doing is setting up the peripheral interface adapter in a parallel interface board inside the Color Computer. It is setting which I/O lines will be outputs and which I/O lines will be inputs. Between lines 25 and 40 you will have to place your own I/O initialization routine. In lines 45 through 90 is the initial screen setup. The four possible commands are printed out on the screen.

This does it for the initialization routine. As we go along, we will see how the current world, target world, delimiter S string, and program line string will be used.

The next section of the program is the initial input section, which takes operator input from the first screen that was shown at the bottom of the initialization procedure. This initial input section is shown in Fig. 4-8. Line number 100 is a basic input statement using the variable C. C will be looking for the numbers 1 through 4, delineating either local, remote, or program modes, or exit. After the input is received, the next line, line number 105, will clear the screen once again. Line 110 will then jump to one of four locations depending on the number stored in variable C. If the number selected is 1, or local mode, the program will then jump to line number 300 and resume operation. Remote mode will

```
45 PRINT"SELECT ONE OF THE FOLLOWING"
50 PRINT
60 PRINT"1.LOCAL MODE"
70 PRINT"2.REMOTE MODE"
80 PRINT"3.PROGRAM MODE"
90 PRINT "4.EXIT"
100 INPUT C
105 CLS
110 ON C GOTO 300,500,600,700
120 CLS
130 GOTO 40
```

Fig. 4-8. Listing of initial-input routine.

```
300 REM
301 REM *LOCAL MODE HANDLER*
302 REM
303 PRINT @ 480, "*";
305 SOUND 210,2:INPUT C$
310 IF C$=""THEN GOTO 303
312 IF C$="Q" THEN CLS:GOTO 40
315 GOSUB 330
320 GOTO 300
322 IF E=1 THEN GOTO 303
327 GOSUB 330
328 GOTO 300
```

Fig. 4-9. Listing of local mode handler.

select line 500, program mode will select line 600, and exiting the program or selecting number four will send the program to line 700.

Let's follow the program as it would go should you select mode number one or the local mode. Assuming that from line 110, we now jump to 300, the local mode handler is shown in its entirety in Fig. 4-9. After the title lines, line number 303, realizing that the screen is now cleared from before we got to the handler, will print an asterisk (*) at the bottom line of the display. It will, on line 305, sound a high-pitched beep, and then enter into a string input mode. The * serves as a prompt for operator input. It assumes that you know the list of commands that are available.

After the operator has input the command, line 310 checks to see if a valid, or any command at all, has been entered. Simply pressing the return key and not entering any ASCII data will cause the program to wrap back to line 303, which will simply repeat the prompt. If operator input was an ASCII Q, it would then jump all the way back to the beginning of the program where the first screen is displayed. This is the first of many QUIT possibilities. This allows you to get out of a program fast if you've made a selection mistake.

If you get through the no-input and Q tests, immediately line 315 jumps to a subroutine located at line 330. This subroutine is the check-entry-against-table subroutine. It will look at what you did type and determine if it is a valid command. This subroutine is shown in Fig. 4-10. Let's go over it now.

Realizing what the operator has typed is in variable C$, the first instruction after the title lines is located at line 341. This instruction will actually search through C$ looking for the first occurrence of the ASCII (.) delimiter. The commands in the command coordinator program must be terminated with a period. Therefore, if you were to select the HOME command, HOME would have to be: HOME. without the period (.) an error will be indicated.

```
330 REM
335 REM *CHECK ENTRY AGAINST TABLE*
340 REM
341 N=INSTR(1,C$,S$)
342 IF N=0 THEN E=1:GOTO 410
343 W$=LEFT$(C$,N-1)
345 IF W$="EDIT" THEN GOSUB 1000
350 IF W$="EXECUTE"THEN GOSUB 1500
355 IF W$="GOTO" THEN GOSUB 2000
360 IF W$="GET" THEN GOSUB 2500
365 IF W$="HOME" THEN GOSUB 3000
370 IF W$="LIST" THEN GOSUB 3500
375 IF W$="MOVE" THEN GOSUB 4000
380 IF W$="PLACE" THEN GOSUB 4500
385 IF W$="PLOT" THEN GOSUB 5000
390 IF W$="PROGRAM" THEN GOSUB 5500
395 IF W$="RUN" THEN GOSUB 6000
400 IF W$="TARGET" THEN GOSUB 6500
410 IF E=1 THEN GOSUB 220
415 IF E=2 THEN GOSUB 250
420 E=1:RETURN
```

Fig. 4-10. Listing of check entry against table.

This line, 341, will select the number of locations into the string that the period has been found. The next line, 342, will check to see if a period was indeed found. If no period was found, the error indicator is set and you are sent on to an error subroutine. We'll go over this later. Assuming that there was a period detected, line 343 uses an instruction that extracts all data entered prior to the period, or the actual command itself, and moves it into another variable called W$. From there lines 345 through 400, W$ is checked against all of the valid commands. When a valid command is found, its particular subroutine is jumped to. If no valid command is found, the error variable E is to be set to a 1 from initialization. In line 410, if E is a 1, then you must go to a subroutine that indicates this error. If E is a 2, another subroutine is suggested. We will go over these error subroutines next. If there is no error detected, the initial condition is again set and the program returns to the Local mode handler.

Going back to Fig. 4-9, after line 327 or the place where the check-entry-against-table subroutine is, the next line is a simple GOTO to the beginning of the local mode handler. So commands are entered with the local mode handler and checked against the table in the check-entry-table subroutine; the command is extracted and either executed or flagged as an error and then the whole program wraps back and asks for another command.

Let's get on to those errors. The error-handling routines

```
200 REM
210 REM *ERROR HANDLING ROUTINES*
215 REM
220 SOUND 10,5:PRINT"[INVALID COMMAND]"
230 RETURN
250 SOUND 10,5:PRINT"[BAD INPUT VALUE]"
260 RETURN
270 REM
```

Fig. 4-11. Listing of error handler.

are in line numbers 200 through 270. Fig. 4-11 shows this section of the program. The error trap at location 220 is the invalid command, or if E = 1, it will be called. It simply sounds a low-frequency tone for about a second, prints INVALID COMMAND and then returns. The second one is if E = 2 and that will show a BAD INPUT VALUE. We'll get into the exact reasons for a bad input value later. A third error condition is encountered when you break a program running through the use of pushing a key on the keyboard. This will go into a breakpoint routine that will be explained later when we go over the run routine.

Before we go any further let me explain the way in which a routine must be entered in from the keyboard or from a remote port. As said earlier, the English command word must be terminated in a period. It is very possible that that word may require two variables afterwards. In the case of the MOVE, GET, or PLACE commands those variables would be the X and Y locations. Periods (.) must be entered after each one of those. A valid command would then be MOVE 4.8. In this particular command the table is being commanded to move to location X of 4 and Y of 8. Each one of these periods serves as a delimiter for the program logic. As we get into the various subroutines, you will see how they are used.

Now that we've covered the local mode, later we'll go into each instruction and how they work. Let's jump back to the initial screen and go down to selection number three, the program mode. The reason I am skipping over selection two is because it becomes very machine specific.

Selection number three, the program mode, will allow you to enter in up to 50 separate commands into a program buffer. These commands would be the same as in the table that the local mode handler checks. By selecting number 3, a second menu appears on the screen. Fig. 4-12 is a representation of this screen.

From here, you once again have five choices. The first choice is to enter a program in from the keyboard, second is to edit the program, third is to list, fourth is to actually run the program, and the fifth is to return to the first screen's menu. Let's go into the actual program logic that is used to run this menu screen and enter your selections.

The line numbers for the program mode handler are shown in Fig. 4-13. Here, after the title lines of 600 to 610, we see from line 615 to line 645 the menu actually being printed on the screen. Realize that after this initial screen was selected or printed, the screen was blanked after the first selection. So

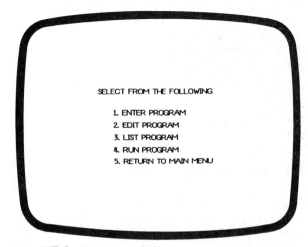

Fig. 4-12. Representation of second program screen prompts.

```
600 REM
605 REM *PROGRAM MODE HANDLER*
610 REM
615 PRINT"SELECT FROM THE FOLLOWING"
620 PRINT
625 PRINT"1. ENTER PROGRAM"
630 PRINT"2. EDIT PROGRAM"
635 PRINT"3. LIST PROGRAM"
640 PRINT"4. RUN PROGRAM"
645 PRINT"5. RETURN TO MAIN MENU"
650 INPUT A
655 IF A=5 THEN CLS:GOTO 40
660 ON A GOSUB 5500,1000,3500,6000
665 CLS:GOTO 615
700 POKE 65407,51:END
```

Fig. 4-13. Listing of program-mode handler.

you would have a clean screen that this second menu comes up on. Once again, in line 650, the selection is looked at in a variable called A. A is looking for the numbers 1 through 5. Line 655 will actually terminate the program mode handler if selection number 5 is made by the operator. It will also clear the screen. If it is not 5, meaning that it is between 1 and 4, line 660 comes into play. Once again, here is a multiple-statement-check procedure that looks at the value of variable A and, this time, will go to a subroutine depending on which number is selected. The enter program mode will jump to a subroutine located at line 5500, the edit program selection will jump to an edit routine located at line 1000; the list routine is located at line 3500, and the run program routine at line 6000.

Let's assume the operator is entering the program; therefore, at this time, let's GOSUB to line 5500. Fig. 4-14 shows the enter program subroutine. As you can see from the figure, the first thing this handler does is to print on the screen a question, whether this a new program or not. Your input, at line 5520, will be a Y or an N indicating a yes or a no answer. If it is a no answer, meaning it is not a new program, you will

```
5500 REM
5505 REM *PROGRAM INPUT HANDLER*
5510 REM
5515 PRINT@480,"NEW PROGRAM";
5520 INPUT A$
5525 IF A$="N" THEN GOTO 1000
5530 IF A$="Q" THEN E=0:RETURN
5535 FOR PC=1 TO 50
5540 PRINT@480," >";PC;" ";
5545 SOUND 210,2:INPUT PI$
5550 IF PI$="" THEN GOTO 5540
5555 IF PI$="Q" THEN E=0:RETURN
5560 PL$(PC)=PI$
5565 NEXT PC
5570 PRINT@480,"MEMORY FULL":SOUND 10,5
5575 E=0:RETURN
```

Fig. 4-14. Listing of program-input handler.

basically be editing an old one. Then, if you'll notice in line 5525, the program, upon receiving this N, will go to the edit subroutine and carry out the program from there. If it is not an N, perhaps it is a Q. Line 5530, once again, allows us to quit what we might have accidentally selected. Line 5535 assumes that you have indeed typed in a yes, meaning you are really entering a new program. Here a program counter, or variable PC, is set up to allow for 50 program lines. Remember that in the initialization phase we allowed for 51. There is a reason for this odd number, and we'll get to that later.

Line 5535 now sets up our program counter. It starts the program counter at 1. The next line, 5540, will print a greater than sign (>), then the current value of the program counter and a space. Line 5545 will give the prompt sound, which is a high-pitched half-second tone (beep) and then look for an input. This input is now being stored into PI$, or program input variable. If in line 5550 nothing has been entered or, simply, the enter button has been pushed, it will prompt you with the same program counter number.

Once again in line 5555, the Q option of quitting is allowed. Line 5560, assuming that you have now finished entering in a real command that was not an ENTER or a Q, will now put what you entered from the keyboard of PI$ into the program line $ array pointed to by the program counter. Now let's digress a little to explain this. In the beginning, a dimension statement was set up to where PL$ was allowed to handle 51 entries. Each entry will be a string of ASCII characters delimited by a carriage return. What you have done in line 5545 is to enter a string of characters terminated by a carriage return. Line 5560 will now put what you have just entered into the first location of that earlier dimension array. Line 5565, saying NEXT PC, will increment the value of the program counter to 2. The program now jumps back to line 5540; gives another prompt with the new PC, which is 2, a space; and asks for the next program line.

As you move through, the program counter will increment until it gets to number 50. After 50 is entered, the program will fall through to line 5570 where a message will be printed, at the bottom of the screen, saying that the memory is now full and a lower tone beep of about 1 second will be heard. At that point in time, the program input handler is terminated and it returns back to the program mode second screen menu. At any time during the entry of the program you can stop incrementing line numbers by simply typing a Q.

If we were to have selected the edit program, we would have been sent to a subroutine located at line number 1000. It is time to go into this subroutine, which is shown in Fig. 4-15. The EDIT program routine basically begins at line 1015. Here, another question is asked on the screen. At the

```
1000 REM
1005 REM *EDIT PROGRAM ROUTINE*
1010 REM
1015 PRINT@480, "LINE NUMBER";
1020 SOUND 210,2: INPUT PC$
1025 IF PC$="Q" THEN RETURN
1030 IF PC$="" THEN GOTO 1015
1035 PC=VAL(PC$)
1040 EL$=PL$(PC)
1045 PRINT@480,EL$
1050 PRINT@480," >";PC;" ";
1055 SOUND 210,2:INPUT EL$
1060 IF EL$="Q" THEN RETURN
1065 IF EL$="" THEN GOTO 1050
1070 PL$(PC)=EL$
1075 GOTO 1015
```

Fig. 4-15. Listing of EDIT routine.

bottom of the screen a phrase is printed requesting which line number you would like to edit. The prompt sound is once again repeated, and the input goes into a PC$ variable. Again, this input is checked for either quit or a no entry, and appropriate actions are taken. If it is indeed a valid line number or any kind of number line, 1035 will produce an actual numbered value from this string and call it PC. Line 1040 will then go off into the program line array and select the particular line number that you have entered in from the keyboard and send it to edit line string, or EL$, variable. Line 1045 will print on the screen what is located at that line number. Line 1050 will then, underneath that, print a prompt, the current PC, and a space, and it now allows you to enter a new command into that same line number. This is the way the edit system works in this particular program. The prompt beep is sounded, and the edit line is now entered into the EL$ in line 1055. Again, lines 1060 and 1065 check for a quit or no-entry condition. If an actual command has been entered, it is then stored back in the current PC location in the program line $ array at line 1070. Then, the program loops back up again to line 1015 and asks you for another line number.

You could go on all day changing line numbers. If you were going to them in order, it would be better and smarter to jump to the program input routine; however, you may have three or four specific line numbers you may want to change. How would you know which line numbers you'd want to change without having all of them in front of you? That's where the list program selection comes in handy. Going back to Fig. 4-13, if we had selected number three, LIST program, we would then be sent off to a subroutine located at line 3500. This routine is shown in Fig. 4-16. As you can see, each one of these routines is relatively simple, and the program goes along very quickly.

```
3500 REM
3505 REM *LIST PROGRAM ROUTINE*
3510 REM
3515 PRINT@480,"PRINT LISTING";
3520 INPUT A$
3525 IF A$="Y" THEN GOTO 3575
3530 FOR PC=1 TO 51 STEP 3
3535 FOR FC=1 TO 3
3540 PRINT@480,PC+(FC-1);" ";PL$(PC+(FC-1))
3545 NEXT FC
3550 INPUT A$:SOUND 210,2
3555 IF A$="Q" THEN E=0:RETURN
3560 NEXT PC
3565 E=0
3570 RETURN
3575 PRINT@480,"PRINTER READY";
3580 INPUT A$
3585 IF A$="N" THEN GOTO 3575
3590 IF A$="Q" THEN E=0:RETURN
3595 PRINT@480, "PROGRAM NAME"
3600 INPUT N$
3605 PRINT#-2,N$
3610 PRINT#-2
3615 FOR PC=1 TO 50
3620 PRINT#-2,PC;" ";PL$(PC)
3625 IF PL$(PC)="END." THEN E=0:RETURN
3630 NEXT PC
3635 PRINT#-2:E=0:RETURN
```

Fig. 4-16. Listing of LIST routine.

Lines 3500 and 3510 are simply the title space showing the fact that it is the LIST routine. From there it puts a question up on the screen asking you if you want to print the listing. This is done on line 3515. Once again, it has asked for a yes or no answer. Line 3525 checks to see if there is a yes; then it will go to the actual print routine in 3575. Assuming that it is not being printed, it will be listed on the screen. Line 3530 is the beginning of the screen listing portion of the program. What this does is set up the program counter. Notice here that the program counter is set from 1 to 51 with something called step 3. The step 3 will mean that it will go from 1 to 3 to 6 to 9, etc. The reason 51 was chosen is because it is easily divisible by 3.

The next line sets another counter called FC and that is 1 to 3. What this will basically do is from line 3530 to line 3545, it will print out three lines of program from the program line array stored when you were entering the program from the keyboard. Line 3540 is where the actual printing happens. PC is the global outside counter; FC is an internal 1 to 3 counter. The action on the screen will be three lines printed, and then in line 3550, it will ask you to enter in a QUIT command or just hit ENTER to get three more lines printed. If the QUIT command is entered, it will be picked up in line 3555 and the program will return to the main program mode menu. If it is not quit, as you can see, line 3560 will get the next global 3 lines of code. FC will print it out on the screen.

If you had selected a yes when it asked you if you wanted to print the listing, then line 3575 asks you if the printer is ready. This means if the printer is selected, connected, and possibly turned on. If you enter in a no, then it will ask you again and line 3585 will send you back up to the question "printer ready?" If in line 3580 a Y is encountered, it will jump down to line 3595 and ask you what name you wish to give to the listing. At this time the name goes into N$ in line 3600. From there it is printed out to the print device.

Now in this listing, line 3605 shows a statement called PRINT#-2. In the MicroSoft BASIC used with the Color Computer this is the printer output. Your system may require LPRINT or some other such code. As you can see in line 3605, N$ is sent out to the printer. Line 3610 will then skip a line so that there is a line between the actual listing and the name of the program (basically for aesthetic reasons). Line 3615, once again, sets up the program counter from 1 to 50. Line 3620 will then print out a line from the program line string array. The format of this printout is the program counter number, a space, and then the line.

Line 3625 will check to see if the command just printed is an END command. If it is an END, printing will stop and return you back to the main program mode input screen. If it is not an END command, 3630 with NEXT PC, will start the process up with PC 2, then 3, then 4, and so on until 50. When 50 is done, it automatically jumps back to the program screen.

So that's it for the list program. Looking back at Fig. 4-13 there are other choices. One other is to run the program, which is where things get a little more complex. Running a program will actually result in executing each one of the valid commands. We will start off with a description of the run program handler and then eventually get into each subroutine, depending on the instruction that is called.

Fig. 4-17 shows the RUN routine in its entirety. It is a very short subroutine but it invokes most other subroutines within the whole program. Lines 6000 through 6010 are the title statement. Line 6020 sets up the program counter. Line 6025 and 6030 are two program lines that will strobe the keyboard looking for a keypress. This is not like an input statement where the keypress must be ended by a carriage return. This is looking for a key-down situation. Your computer may use a different type of mechanism than the INKEY$ statement. As I said, basically what it does is look for a key down. This is the abort mechanism that was mentioned earlier in the

```
6000 REM
6005 REM *RUN PROGRAM ROUTINE*
6010 REM
6020 FOR PC=1 TO 50
6025 A$=INKEY$
6030 IF A$ <> "" THEN E=3:RETURN
6035 C$=PL$(PC)
6040 PRINT@480,PC;" ";C$
6045 IF C$="END." THEN GOTO 6065
6050 GOSUB 330
6055 IF G=1 THEN G=0:GOTO 6035
6060 NEXT PC
6065 PRINT@480, "PROGRAM STOP":SOUND 10,10
6070 E=0:RETURN
6075 REM
```

Fig. 4-17. Listing of RUN routine.

```
2000 REM
2005 REM *GOTO ROUTINE*
2010 REM
2015 GX=INSTR(N+1,C$,S$)
2020 IF GX=0 THEN E=2:RETURN
2025 GN$=MID$(C$,N+1)
2030 GN=VAL(GN$)
2035 PC=GN
2040 G=1
2045 E=0
2050 RETURN
```

Fig. 4-18. Listing of GOTO routine.

chapter as an emergency measure. Should the program run amuck, you have the possibility of stopping immediately by pressing any key on the keyboard. In this implementation, if there is a key down detected, the program will give an error condition of three in the E variable and then return. If you remember in the discussions of the error-handling routines, an error of three basically tells the system, and will put up on the screen, that a manual break has occurred.

Line 6035 goes on from line 6020 if no key is detected. Line 6025 will get the first program line in the program line array pointed to by the program counter set up at line 6020. Notice that it brings this program line into the variable C$. C$ is the variable that is used in the local mode input for running or executing programs in real time. There is a significance in using C$, and you'll see it in a minute.

After pulling it out of the array, the instruction, in its entirety, is printed on the screen along side its line number so that the operator may see the actual program and how it is tracing on the screen. Next line, 6045, checks to see if this is an END statement. An END statement will cause it to go to line 6065 which will print "Program Stop". The program will be stopped, and you will be sent back to the main program mode screen. If it is not an END, it is asking to run a subroutine located at 330. If you will recall this subroutine is the check-entry-against-table routine, which is normally entered from the local mode handler with the variable C$ carrying the actual command. So, from here on out the run program acts like the local mode program except at the end of each local mode execution subroutine, it is brought back to the RUN program routine where in line 6055 a certain test is executed. This test goes along with the GOTO routine. I'll cover this a little later.

As things crank out, the next PC is selected, and it goes through running the program from 1 to 50. At the end of 50, if no END has been detected, the program stop will be in effect and it will go back to the main screen. It's now time to go into the actual command subroutines.

The first one on the list, that I have not gone over so far, is the GOTO routine. It is checked for in line 355 of the check-entry-against-table routine that can be seen back in Fig. 4-10. GOTO goes to a subroutine located at line 2000. This subroutine is shown in Fig. 4-18. The GOTO subroutine will be our first look at a routine that takes an argument out of the program line or the command line after the first period delimiter. This is a good one to start with because it only selects one argument. Other routines will have two. As we go to Fig. 4-18, we see that the program first gets that portion of the command string in line 2015 between the first and second periods. It will only take the ASCII values between the two. It will not take one or the other delimiter with it.

This particular portion of the GOTO routine will be a line number. It will, typically, only be two characters long. If the line number is not found, in other words, there is no argument with the instruction; line 2020 will find that out and return with an error. That error is number two. As I said earlier, it would be explained. This error is called a bad input value. If you, the operator, get a bad-input-value message on the screen, it means that you either forgot a delimiter during an argument or you forgot an argument. Line 2025 is where the actual piece of the command string containing the argument is sectioned off and put into what is called GN$. Line 2030 changes this ASCII input into an actual decimal number with the VALUE statement. This value number is then turned into the new program counter in 2035. Line 2040 sets this variable G or GOTO flag to a one. This will tell the run program not to go to the next following PC but to actually put in this new value of PC and pull the program line from there. If you quickly return back to Fig. 4-17, you will see that in line 6055, if the GOTO flag is one, it resets the GOTO flag and then goes back up to 6035, which will get the new line at the new program counter value instead of going on line 6060, which would simply increment the program value.

After the GOTO flag is set, the error condition is reset and you are returned to the main handler. GOTO works basically just like any program command such as BASIC or any other high-level language. It allows you to skip around in the program from one area to another just as you are seeing it used in this program.

The next program in the line to be discussed is the GET program. GET sends you to a subroutine starting at line 2500. This is the beginning of the actual MOVE command structures. They are all interrelated in that they will all use a

```
2500 REM
2505 REM *GET ROUTINE*
2510 REM
2515 GOSUB 7000
2520 TPX=0:TPY=0:GT=1
2525 IF X=CPX THEN GOTO 2550
2530 IF X<CPX THEN DRX=0:GOTO 2545
2535 IF X>CPX THEN DRX=1
2540 TPX=X-CPX:GOTO 2550
2545 TPX=CPX-X
2550 IF Y=CPY THEN GOTO 2575
2555 IF Y<CPY THEN DRY=0:GOTO 2570
2560 IF Y>CPY THEN DRY=1
2565 TPY=Y-CPY:GOTO 2575
2570 TPY=CPY-Y
2575 GOSUB 7500
2580 E=0:RETURN
```

Fig. 4-19. Listing of GET-BLOCK subroutine.

```
7000 REM
7005 REM *COORDINATE CONVERTER ROUTINE*
7010 REM
7015 NX=INSTR(N+1,C$,S$)
7020 IF NX=0 THEN E=2:RETURN
7025 X$=MID$(C$,N+1,1)
7030 NY=INSTR(NX+1,C$,S$)
7035 IF NY=0 THEN E=2:RETURN
7040 Y$=MID$(C$,NX+1,1)
7045 REM *CONVERT VALUES*
7050 IF X$="1" THEN X=0:GOTO 7095
7055 IF X$="2" THEN X=32:GOTO 7095
7060 IF X$="3" THEN X=64:GOTO 7095
7065 IF X$="4" THEN X=96:GOTO 7095
7070 IF X$="5" THEN X=128:GOTO 7095
7075 IF X$="6" THEN X=160:GOTO 7095
7080 IF X$="7" THEN X=192:GOTO 7095
7085 IF X$="8" THEN X=224:GOTO 7095
7090 E=2:RETURN
7095 IF Y$="1" THEN Y=0:GOTO 7140
7100 IF Y$="2" THEN Y=32:GOTO 7140
7105 IF Y$="3" THEN Y=64:GOTO 7140
7110 IF Y$="4" THEN Y=96:GOTO 7140
7115 IF Y$="5" THEN Y=128:GOTO 7140
7120 IF Y$="6" THEN Y=160:GOTO 7140
7125 IF Y$="7" THEN Y=192:GOTO 7140
7130 IF Y$="8" THEN Y=224:GOTO 7140
7135 E=2:RETURN
7140 E=0:RETURN
```

Fig. 4-20. Listing of coordinate-converter subroutine.

coordinate converter subroutine to actually find the place to GET, PLACE, or MOVE.

We begin the discussion with a description of this coordinate converter routine. The GET subroutine itself is shown in Fig. 4-19. We will be interweaving discussions on this and Fig. 4-20, which is the coordinate converter subroutine.

The first action that takes place in the GET subroutine is that the program is vaulted off to another subroutine located at line 7000. This is the coordinate converter subroutine.

This routine will take the two arguments from the GET statement and turn them into actual step-counter values. This seems an appropriate time to go over that routine. Look at Fig. 4-20. At line numbers 7015 through 7025, the first argument, or the X value, is found and then stored away in a variable called X$. A similar action is done in lines 7030 through 7040 where the Y argument is found and put into the Y$ variable. From 7045 on the values are converted into step counter values, as mentioned before. Between 7050 and 7085 the X value that was taken out of the command statement is checked against each one of the eight possible values. Each of these eight values represents 32 step counts. As you can see by looking at a typical line such as line number 7065, if the X$ value is four then that would mean 96 step counts. If none of these are selected, in other words if you have accidentally put in a nine, a zero, or some greater number, you will get a bad argument error condition as shown in line 7090. Lines 7095 through 7130 do a similar conversion for the Y argument. When the subroutine ends in line 7140, it will go back to the GET routine with an X and a Y value. These will be called the target step values or the target positions.

Going back to the GET routine in Fig. 4-19 after the coordinate converter routine is done, the program will return to line 2520. Here the target values of X and Y are set to zero, and the GET routine delimiter is changed to a one. Between line 2525 and line 2535 relativity between the current position and the target position of the arm is checked. In line 2525 the current X position and the target X position are checked for equality. If they are equal, the program is sent on to check the current Y position against the target Y position. If the X current position is not the same as the target position, line 2530 will check for a less than (<) relativity, 2535 will check for greater than (>). Each one of these lines will set the direction flag to a reverse direction, which is indicated by a zero in variable DR. The positive or forward direction is a one in the DR variable. After it is determined which direction to move the target position register or TPX in, the X value is computed by either using line 2540 or 2545. When you are finished with the X values, similar things happen for the Y values between line 2550 and line 2570.

At this point, with a new TPX and TPY variable, the GET routine sends you to the motion subroutine. The motion subroutine will actually move the arm to the location specified by TPX and TPY. The motion subroutine will also check the value of the GT variable and will, when it gets to the point specified by TPX and TPY, maneuver the Z axis, turn on the gripper, and pull up the part that you are trying to get or the block that you are trying to get. When it finishes with the motion subroutine, it comes back in line 2530, sets the position to zero, and returns.

We will go over the motion subroutine in a little while. Let it be known that the coordinate converter routine is used in all the motion-type routines. The motion subroutine itself in 7500 is called by all of them, also. This is not to be confused

```
4000 REM
4005 REM *MOVE ROUTINE*
4010 REM
4015 GOSUB 7000
4020 TPX=0:TPY=0
4025 IF X=CPX THEN GOTO 4050
4030 IF X < CPX THEN DRX=0:GOTO 4045
4035 IF X > CPX THEN DRX=1
4040 TPX=X-CPX:GOTO 4050
4045 TPX=CPX-X
4050 IF Y=CPY THEN GOTO 4075
4055 IF Y < CPY THEN DRY=0:GOTO 4070
4060 IF Y > CPY THEN DRY=1
4065 TPY=Y-CPY:GOTO 4075
4070 TPY=CPY-Y
4075 GOSUB 7500
4080 E=0:RETURN
```

Fig. 4-21. Listing of MOVE routine.

```
4500 REM
4505 REM *PLACE ROUTINE*
4510 REM
4515 GOSUB 7000
4520 TPX=0:TPY=0:PT=1
4525 IF X=CPX THEN GOTO 4550
4530 IF X < CPX THEN DRX=0:GOTO 4545
4535 IF X > CPX THEN DRX=1
4540 TPX=X-CPX:GOTO 4550
4545 TPX=CPX-X
4550 IF Y=CPY THEN GOTO 4575
4555 IF Y < CPY THEN DRY=0:GOTO 4570
4560 IF Y > CPY THEN DRY=1
4565 TPY=Y-CPY:GOTO 4575
4570 TPY=TPY-Y
4575 GOSUB 7500
4580 E=0:RETURN
```

Fig. 4-22. Listing of PLACE-BLOCK subroutine.

```
3000 REM
3005 REM *HOME ROUTINE*
3010 REM
3015 LX=PEEK(49153)
3020 LX=LX AND 128
3025 IF LX=128 THEN GOTO 3040
3030 POKE 49152,7:POKE 49152,6:POKE 49152,7
3035 GOTO 3015
3040 LY=PEEK(49153)
3045 LY=LY AND 64
3050 IF LY=64 THEN GOTO 3065
3055 POKE 49152,7:POKE 49152,5:POKE 49152,7
3060 GOTO 3040
3065 LZ=PEEK(49155)
3070 LZ=LZ AND 128
3075 IF LZ=128 THEN GOTO 3090
3080 POKE 49152,7:POKE 49152,3:POKE 49152,7
3085 GOTO 3065
3090 E=0:CPX=0:RETURN
```

Fig. 4-23. Listing of HOME routine.

with the MOVE subroutine, which I will be going over next.

MOVE, as shown in Fig. 4-21, simply does what is stated. It does not pick up a block or place a block; it just goes to a location. However, if the gripper has a block before the MOVE instruction, it will hold on to that block throughout the MOVE command. Conversely, if it has no block, it will not pick one up along the way.

Looking at Fig. 4-21, you'll notice a striking resemblance between the MOVE routine and the GET routine. That's because they are almost one and the same except for various line number changes, but the most significant difference is the fact that in line 4020, where the temporary values of TPX and TPY are zeroed out, you'll notice that there is no GT variable stated there. This is because you don't want the gripper status to be changed during a MOVE instruction. Everything else in the MOVE subroutine is the same as in the GET routine. It goes out, in line 4015, to the subroutine in line 7000 to pull the X and Y arguments off of the command string. When it is finished calculating the values, it then goes down to line 4075 where it actually performs the motion by going to the subroutine at line 7500.

The PLACE subroutine, shown in Fig. 4-22, is once again a carbon copy of GET and MOVE. The difference is that the GT variable is now set to zero. This will tell the motion routine in 7500 to release the block after the MOVE is completed. No further description of the PLACE routine should be necessary, except to say that it will place a block or any object at the location X,Y.

The only other motion-type command left to discuss is the HOME command. As you recall, when HOME is implemented, the arm goes back to location 1,1. This is basically like the initialization procedure. In fact, the HOME command is an initialization. It is not handled through the normal motion routines, and it can be seen in Fig. 4-23. Let's go over the operation of this subroutine. It starts at line 3000, and in line 3015 it will check the value of the X-direction limit switch. Between lines 3015 and 3020, the peripheral interface adapter IC used with the Color Computer is checked for low level on this limit switch input. If it is a low level, it will give the value 128 shown in line 3025. These three lines, 3015 through 3025, will probably change depending on your machine-specific application. However, in line 3025, if the limit switch is found, it vaults the program to line 3040, which will start up in the Y direction, but let's assume, for now, that the limit switch is not activated and the arm is somewhere in the middle of its travel. Line 3030 will actually perform one step of the stepper motor. Line 3035 will go back and look at the switch again.

As you can see, between these five lines, you have a loop that will constantly step until the limit switch activation is found. The direction of movement is in the negative, assuming that you are going toward the limit switch. When the X limit switch is detected, the program moves down to line 3040 and between lines 3040 and 3060 the exact same program manipulation occurs looking for the Y direction

```
7500 REM
7505 REM *MOTION ROUTINE*
7510 REM
7515 IF DRX=0 AND GT=1 THEN POKE 49154,
     7 :GOTO 7525
7520 IF DRX=1 AND GT=1 THEN POKE 49154,0
7525 FOR C=1 TO TPX
7530 POKE 49152,7:POKE 49152,6:POKE 49152,7
7535 NEXT C
7540 IF DRX=0 THEN CPX=CPX-TPX:GOTO 7550
7545 CPX=CPX+TPX
7550 IF DRY=0 AND GT=1 THEN POKE 49154,
     7 :GOTO 7560
7555 IF DRY=1 AND GT=1 THEN POKE 49154,0
7560 FOR C=1 TO TPY
7565 POKE 49152,7:POKE 49152,5:POKE 49152,7
7570 NEXT C
7575 IF DRY=0 THEN CPY=CPY-TPY:GOTO 7585
7580 CPY=CPY+TPY
7585 IF GT=1 THEN GOTO 7640
7590 IF PT=0 THEN RETURN
7595 POKE 49154,0
7600 FOR C=1 TO 32
7605 POKE 49152,7:POKE 49152,3:POKE 49152,7
7610 NEXT C
7615 POKE 49154,68
7620 FOR C=1 TO 32
7625 POKE 49152,7:POKE 49152,3:POKE 49152,7
7630 NEXT C
7635 RETURN
7640 POKE 49154,64
7645 FOR C=1 TO 32
7650 POKE 49152,7:POKE 49152,3:POKE 49152,7
7655 NEXT C
7660 POKE 49154,4
7665 FOR C=1 TO 32
7670 POKE 49152,7:POKE 49152,3:POKE 49152,7
7675 NEXT C
7680 RETURN
```

Fig. 4-24. Listing of motion routine.

limit switch and stepping the Y stepper motor. When the Y direction is found, the same happens in the Z direction, which is the gripper arm. From lines 3065 to 3085, the Z direction stepper motor, which moves the gripper up and down, will be looking for a limit switch activation and moving the up direction, which is the reverse direction for that particular motor.

When all three limit switches are activated, the program returns in line 3090. As you can see, this does not use the motion subroutine in line 7500 nor does it use any of the current-position registers or target-position registers. You will notice that when it returns, in line 3090, it sets the current-position CPX variable to zero, and the CPY variable to zero.

Before we go on to any other commands, it is now time to describe the operation of the motion subroutine in line 7500. As you recall, this is the subroutine that actually manipulates the stepper-motor interface, moving it in whichever direction by as many steps as were specified by the calling subroutine. The motion subroutine is shown in Fig. 4-24. Starting off with 7515, after the three title lines, we find that it does a check, between 7515 and 7520, of a combination of directions and gripper status. Line 7515 checks to see if it is a reverse direction with the gripper on. In this case it would be a reverse GET routine. If that is the case, then the parallel output port that directs direction and gripper on/off would be manipulated in the POKE statement of this line to satisfy the conditions of DRX and GT. The same is true of the opposite direction of line 7520.

We make a short digression to Fig. 4-25, which is a programmer's model of the two 8-bit output ports for the parallel interface to my particular stepper-motor system. Looking at the figure, at the top is the A port, which basically controls the enable of the steppers and the actual stepping. The bottom portion of the figure is the B port, which is the gripper status, flasher status, beeper status, and the direction of each of the three stepper motors. The program line in Fig. 4-24, line 7515, is looking for a reverse direction gripper on in the B port, which is denoted by location 49154 memory. Looking back at the B port, a value of seven, which is being POKEd to that port would turn on, in the reverse direction, all X, Y, and Z stepper motors and it would turn on the gripper, the flasher, and the beeper. This is necessary because GT equaling one refers to this as a GET instruction. DRX equals zero refers to the reverse direction. After that, the program line goes to line 7525. Line 7525 sets a counter from 1 to the TPX value that comes in from the calling subroutine. If you recall, this TPX value will be the amount of steps necessary to step in a particular direction. The following line, 7530, does the actual step. Line 7535 will increment the counter, C, to perform the many steps required. At the end of the stepping action, the current position register of CPX will be updated depending on the direction you just stepped in line 7540 or line 7545.

Going back to line 7520, if the motion were to be in the forward direction, then a value of 0 would set all the direction bits in the B port register, Fig. 4-25, to the forward direction and, of course, the gripper, flasher and beeper would still be enabled. The same is true for the Y direction starting at line 7550 and going through to line 7580. It is at line 7585 that the Z direction or the gripper motor is encountered. Here, between lines 7585 and 7590 it is looking to see if this is a MOVE routine, a GET routine, or a PLACE routine. A GET routine would be signified by a one in the GT variable. If this were the case, then the program would go from 7585 where it checked for the one to line 7640. Line 7640, going down the figure, is another POKE to the B register. This would turn on the gripper and turn on the forward direction. It then would step down, in lines 7645, 50 and 55, 32 positions that, in my particular case, are the maximum direction down. When it got down there at line 7660, it would then change direction for the next 32 steps shown in lines 7665 and 7670, which brings the block up with the gripper, realizing that the gripper

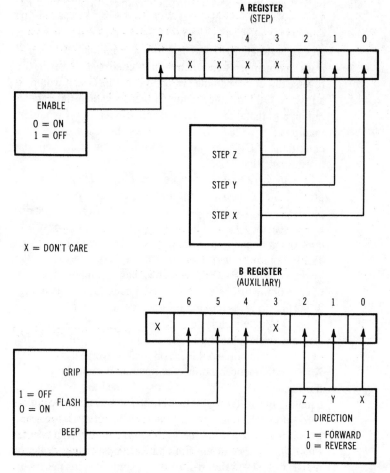

Fig. 4-25. Programmer's model of interface memory locations.

status has been on since the beginning and has shut off. So at the end of line 7675, we now have a block in the gripper, in the upper direction, at the X,Y location that was requested.

Going back to line 7590, if GT was not one, the program then checks to see if it was a PLACE operation. A PLACE operation would be the variable PT equaling 1. If it equals zero, then a MOVE operation is indicated and no further action of the Z axis should be made. Therefore, in line 7590, if it is a MOVE instruction (PT = 0), the program will return. If the PLACE instruction were implemented (PT = 1), the program would go on to line 7595 where a POKE to the B port would signify the status as staying with the grip and moving in the forward direction (that's what the zero stands for). Lines 7600-7610 will move the gripper down to contact an area. Line 7615 would then change the status of the direction and the gripper. Remembering that this is a PLACE instruction, the gripper is now released and from line 7620 through 7630, the arm is then raised again. Line 7635 returns from the subroutine.

The motion routine, as suggested earlier, is very machine dependent. It is my hope that, between the discussions of Fig. 4-24 and the programmer's model of Fig. 4-25, it will be easy for you to see what is going on in this subroutine so that you can make any changes necessary in order to make it work for your system.

Well, there are two other instructions that we have not touched on as yet. They are the PLOT and TARGET routines. These are higher-level routines used to work with the EXECUTE routine, which is not implemented in this listing. PLOT is used to state the current value of the world, being where each work-space section and which block is placed in that section. The target world is a rearrangement of the current world. Recalling from the final assembly robot that we talked about earlier, the current world may be where the radio parts are currently located in their bins. The target world would be where you would like them to end up after assembly. The EXECUTE routine would do the assembly itself.

There are many ways in which to enter in a current world. Let's look at the way that I have implemented the current world entry routine called the PLOT routine, shown in Fig. 4-26. The PLOT routine starts at program line 5000. The first three lines are title lines. From line 5015 through 5065 the program exists. As you will find, it is a very simple entry-type routine. It will allow the operator, through the use of the keyboard, to enter in types of entities that are on the work-space squares or the checkerboard squares or whatever you want to call them. Line 5015 starts off setting up the loop. The loop is for the X direction from one to eight. Line 5020 sets up an internal loop of the Y direction. Let me explain what the two of these lines will accomplish. The lines between 5020 and 5045 will be executed eight times for every one increment of the X value of line 5015. The operator will see, prompted on the screen, at line 5025, the first message being: 1,1 and then -. What that tells you is that it is X-1, Y-1. The operator will then be prompted to enter in an ASCII value of what is in that workspace. In my program the ASCII value will be either a W for a white block or a B for a black block. After that value is entered, the program will check to see if it is a null entry or a quit. A null entry is where no value is typed, just a carriage return was hit. In this case, the program works a little differently than the rest of the program has so far. If you recall in other routines, the null value will reprompt you at the same value. In this application a null value will assume that you are skipping over that block and not putting anything in it nor changing the current information. Therefore, when a null indication happens in line 5030, line 5035 will then pull you out of the routine and down to where you will increment to the next value. After 1,1 the next value prompted will be an X value of one and a Y value of two. It will increment the Y value from one through eight, and then increment the X value to two and repeat the Y values one through eight. As you can see the Y value counter, or the FOR loop, in line 5020 will increment eight times for every one time of line 5015. Line 5045 is where the actual storage of the values goes into the current world array. If you'll recall, we set up the array in the initialization phase of the program. At the end, when all 64 workspace squares have been filled or their values have been entered, line 5060 will prompt the operator with a PLOT FINISHED message. At that point it will return to the calling program. At any time you can stop the prompting by typing a Q and then an ENTER.

```
5000 REM
5005 REM *PLOT ROUTINE*
5010 REM
5015 FOR X=1 TO 8
5020 FOR Y=1 TO 8
5025 PRINT@480,"X";X;" Y";Y;" -";
5030 INPUT A$:SOUND 210,2
5035 IF A$="" THEN GOTO 5050
5040 IF A$="Q" THEN E=0:RETURN
5045 CW$(X,Y)=A$
5050 NEXT Y
5055 NEXT X
5060 PRINT@48),"[PLOT FINISHED]":SOUND 10,10
5065 E=0:RETURN
```

Fig. 4-26. Listing of PLOT routine.

The target world routine, shown in Fig. 4-27, is very similar to the PLOT routine. The only difference being the target world string in line 6545. The rest of the program, except for in line 6560, the prompt TARGET FINISHED message will be shown. As mentioned earlier, the target routine will fill up an imaginary world that you would like the state of the workspace to be following a MOVE session. The target world array and this whole entry routine are not necessary for the operation of the normal program shown here. It will simply be used if the EXECUTE routine is designed. Later, we will go into some ideas of how to design an EXECUTE routine, which borders, basically, on the ideas of artificial intelligence.

```
6500 REM
650  REM *TARGET ROUTINE*
6510 REM
6515 FOR X=1 TO 8
6520 FOR Y=1 TO 8
6525 PRINT@480,"X";X;" Y";Y;" -";
6530 INPUT A$:SOUND 210,2
6535 IF A$="" THEN GOTO 6550
6540 IF A$="Q" THEN E=0:RETURN
6545 CW$(X,Y)=A$
6550 NEXT Y
6555 NEXT X
6560 PRINT@480,"[TARGET FINISHED]":SOUND 10,10
6565 E=0:RETURN
```

Fig. 4-27. Listing of TARGET routine.

From here, it would be helpful to see the entire coordinate controller listing to get a feel for the interrelationships between the routines and to facilitate your being able to enter them into your system. The entire listing for this first program is shown in Fig. 4-28.

XY TABLE SIMULATION

Just as the title suggests, it is possible to totally simulate the action of the XY table we've been designing in the past section of this chapter. For those of you with no mechanical ability or who do not have the desire to actually implement the mechanical portions of the XY table, I have included the design of a computer graphics oriented XY table simulator. The reason for this simulator was to allow remote control of my XY table from another location. With it I can see, in almost real time, the operations of the table on a computer graphics representation, on a television screen, while the actual XY table is in motion. Each one of these applications runs on a separate computer. The simulation computer talks to the coordinate controller computer via a serial RS232 communications link. Commands that are sent to the simulation screen are simultaneously sent to the coordinate controller computer for actual actuation. The simulation computer used was a Texas Instruments 99/4A expanded model. The program that I will be describing runs in Extended BASIC and utilizes an assembly-language routine for scrolling text on the screen. As I go into each routine, I will try to explain adaptations to other computer languages.

```
2 POKE 65407,0
5 REM
6 REM **COORDINATE CONTROLLER**
7 REM **c MARK J. ROBILLARD 1983
8 REM
10 CLS
15 DIM CW$(8,8):DIM TW$(8,8)
20 S$=".":E=1:DIM PL$(51)
25 POKE 49153,9:POKE 49155,9
30 POKE 49152,255:POKE 49154,255
35 POKE 49153,13:POKE 49155,13
40 POKE 49152,255:POKE 49154,255
45 PRINT"SELECT ONE OF THE FOLLOWING"
50 PRINT
60 PRINT"1. LOCAL MODE"
70 PRINT"2. REMOTE MODE"
80 PRINT"3. PROGRAM MODE"
90 PRINT"4. EXIT"
100 INPUT C
105 CLS
110 ON C GOTO 300,500,600,700
120 CLS
130 GOTO 40
200 REM
210 REM *ERROR HANDLING ROUTINES*
215 REM
220 SOUND 10,5:PRINT"[INVALID COMMAND]"
230 RETURN
250 SOUND 10,5:PRINT"[BAD INPUT VALUE]"
260 RETURN
270 REM
300 REM
301 REM *LOCAL MODE HANDLER*
302 REM
303 PRINT @ 480,"*";
305 SOUND 210,2:INPUT C$
310 IF C$="" THEN GOTO 303
312 IF C$="Q" THEN CLS:GOTO 40
315 GOSUB 330
320 GOTO 300
322 IF E=1 THEN GOTO 303
327 GOSUB 330
328 GOTO 300
330 REM
335 REM *CHECK ENTRY AGAINST TABLE*
340 REM
341 N=INSTR(1,C$,S$)
342 IF N=0 THEN E=1:GOTO 410
343 W$=LEFT$(C$,N-1)
345 IF W$="EDIT" THEN GOSUB 1000
350 IF W$="EXECUTE"THEN GOSUB 1500
355 IF W$="GOTO" THEN GOSUB 2000
360 IF W$="GET" THEN GOSUB 2500
365 IF W$="HOME" THEN GOSUB 3000
370 IF W$="LIST" THEN GOSUB 3500
375 IF W$="MOVE" THEN GOSUB 4000
380 IF W$="PLACE" THEN GOSUB 4500
385 IF W$="PLOT" THEN GOSUB 5000
390 IF W$="PROGRAM" THEN GOSUB 5500
395 IF W$="RUN" THEN GOSUB 6000
400 IF W$="TARGET" THEN GOSUB 6500
410 IF E=1 THEN GOSUB 220
415 IF E=2 THEN GOSUB 250
420 E=1:RETURN
425 REM
500 REM
505 REM * REMOTE MODE HANDLER*
510 REM
515 PRINT@480,"[NOT IMPLEMENTED]"
520 SOUND 10,5
530 E=0:RETURN
600 REM
605 REM *PROGRAM MODE HANDLER*
610 REM
615 PRINT"SELECT FROM THE FOLLOWING"
620 PRINT
625 PRINT"1. ENTER PROGRAM"
630 PRINT"2. EDIT PROGRAM"
635 PRINT"3. LIST PROGRAM"
640 PRINT"4. RUN PROGRAM"
645 PRINT"5. RETURN TO MAIN MENU"
650 INPUT A
655 IF A=5 THEN CLS:GOTO 40
660 ON A GOSUB 5500,1000,3500,6000
665 CLS:GOTO 615
700 POKE 65407,51:END
1000 REM
1005 REM *EDIT PROGRAM ROUTINE*
1010 REM
1015 PRINT@480,"LINE NUMBER";
1020 SOUND 210,2: INPUT PC$
1025 IF PC$="Q" THEN RETURN
1030 IF PC$="" THEN GOTO 1015
1035 PC=VAL(PC$)
1040 EL$=PL$(PC)
1045 PRINT@480,EL$
1050 PRINT@480," > ";PC;" ";
1055 SOUND 210,2:INPUT EL$
1060 IF EL$="Q" THEN RETURN
1065 IF EL$="" THEN GOTO 1050
1070 PL$(PC)=EL$
1075 GOTO 1015
1500 REM
1505 REM *EXECUTE ROUTINE*
1510 REM
1515 PRINT@480,"[NOT IMPLEMENTED]":SOUND 10,5
1520 E=0
1525 RETURN
2000 REM
2005 REM *GOTO ROUTINE*
2010 REM
2015 GX=INSTR(N+1,C$,S$)
2020 IF GX=0 THEN E=2:RETURN
2025 GN$=MID$(C$,N+1)
2030 GN=VAL(GN$)
2035 PC=GN
2040 G=1
2045 E=0
2050 RETURN
2500 REM
2505 REM *GET ROUTINE*
```

Fig. 4-28 Entire listing of coordinate controller.

Fig. 4-28. Cont.

```
2510 REM
2515 GOSUB 7000
2520 TPX=0:TPY=0:GT=1
2525 IF X=CPX THEN GOTO 2550
2530 IF X < CPX THEN DRX=0:GOTO 2545
2535 IF X > CPX THEN DRX=1
2540 TPX=X-CPX:GOTO 2550
2545 TPX=CPX-X
2550 IF Y=CPY THEN GOTO 2575
2555 IF Y < CPY THEN DRY=0:GOTO 2570
2560 IF Y > CPY THEN DRY=1
2565 TPY=Y-CPY:GOTO 2575
2570 TPY=CPY-Y
2575 GOSUB 7500
2580 E=0:RETURN
3000 REM
3005 REM *HOME ROUTINE*
3010 REM
3015 LX=PEEK(49153)
3020 LX=LX AND 128
3025 IF LX=128 THEN GOTO 3040
3030 POKE 49152,7:POKE 49152,6:POKE 49152,7
3035 GOTO 3015
3040 LY=PEEK(49153)
3045 LY=LY AND 64
3050 IF LY=64 THEN GOTO 3065
3055 POKE 49152,7:POKE 49152,5:POKE 49152,7
3060 GOTO 3040
3065 LZ=PEEK(49155)
3070 LZ=LZ AND 128
3075 IF LZ=128 THEN GOTO 3090
3080 POKE 49152,7:POKE 49152,3:POKE 49152,7
3085 GOTO 3065
3090 E=0:CPX=0:CPY=0:RETURN
3500 REM
3505 REM *LIST PROGRAM ROUTINE*
3510 REM
3515 PRINT@480,"PRING LISTING";
3520 INPUT A$
3525 IF A$="Y" THEN GOTO 3575
3530 FOR PC=1 TO 51 STEP 3
3535 FOR FC=1 TO 3
3540 PRINT@480,PC+(FC-1);" ";PL$(PC+(FC-1))
3545 NEXT FC
3550 INPUT A$:SOUND 210,2
3555 IF A$="Q" THEN E=0:RETURN
3560 NEXT PC
3565 E=0
3570 RETURN
3575 PRINT@480,"PRINTER READY";
3580 INPUT A$
3585 IF A$="N" THEN GOTO 3575
3590 IF A$="Q" THEN E=0:RETURN
3595 PRINT@480,"PROGRAM NAME"
3600 INPUT N$
3605 PRINT#-2,N$
3610 PRINT#-2
3615 FOR PC=1 TO 50
3620 PRINT#-2,PC;" ";PL$(PC)
3625 IF PL$(PC)="END." THEN E=0:RETURN
3630 NEXT PC
3635 PRINT#-2:E=0:RETURN
4000 REM
4005 REM *MOVE ROUTINE*
4010 REM
4015 GOSUB 7000
4020 TPX=0:TPY=0
4025 IF X=CPX THEN GOTO 4050
4030 IF X < CPX THEN DRX=0:GOTO 4045
4035 IF X > CPX THEN DRX=1
4040 TPX=X-CPX:GOTO 4050
4045 TPX=CPX-X
4050 IF Y=CPY THEN GOTO 4075
4055 IF Y < CPY THEN DRY=0:GOTO 4070
4060 IF Y > CPY THEN DRY=1
4065 TPY=Y-CPY:GOTO 4075
4070 TPY=CPY-Y
4075 GOSUB 7500
4080 E=0:RETURN
4500 REM
4505 REM *PLACE ROUTINE*
4510 REM
4515 GOSUB 7000
4520 TPX=0:TPY=0:PT=1
4525 IF X=CPX THEN GOTO 4550
4530 IF X < CPX THEN DRX=0:GOTO 4545
4535 IF X > CPX THEN DRX=1
4540 TPX=X-CPX:GOTO 4550
4545 TPX=CPX-X
4550 IF Y=CPY THEN GOTO 4575
4555 IF Y < CPY THEN DRY=0:GOTO 4570
4560 IF Y > CPY THEN DRY=1
4565 TPY=Y-CPY:GOTO 4575
4570 TPY=TPY-Y
4575 GOSUB 7500
4580 E=0:RETURN
5000 REM
5005 REM *PLOT ROUTINE*
5010 REM
5015 FOR X=1 TO 8
5020 FOR Y=1 TO 8
5025 PRINT@480,"X";X;" Y";Y" -";
5030 INPUT A$:SOUND 210,2
5035 IF A$="" THEN GOTO 5050
5040 IF A$="Q" THEN E=0:RETURN
5045 CW$(X,Y)=A$
5050 NEXT Y
5055 NEXT X
5060 PRINT@480,"[PLOT FINISHED]":SOUND 10,10
5065 E=0:RETURN
5500 REM
5505 REM *PROGRAM INPUT HANDLER*
5510 REM
5515 PRINT@480,"NEW PROGRAM";
5520 INPUT A$
5525 IF A$="N" THEN GOTO 1000
5530 IF A$="Q" THEN E=0:RETURN
5535 FOR PC=1 TO 50
5540 PRINT@480," >";PC;" ";
5545 SOUND 210,2:INPUT PI$
5550 IF PI$="" THEN GOTO 5540
```

Fig. 4-28. Cont.

```
5555 IF PI$="Q" THEN E=0:RETURN
5560 PL$(PC)=PI$
5565 NEXT PC
5570 PRINT@480,"MEMORY FULL":SOUND 10,5
5575 E=0:RETURN
6000 REM
6005 REM *RUN PROGRAM ROUTINE*
6010 REM
6020 FOR PC=1 TO 50
6025 A$=INKEY$
6030 IF A$ <> "" THEN E=3:RETURN
6035 C$=PL$(PC)
6040 PRINT@480,PC;" ";C$
6045 IF C$="END." THEN GOTO 6065
6050 GOSUB 330
6055 IF G=1 THEN G=0:GOTO 6035
6060 NEXT PC
6065 PRINT@480,"PROGRAM STOP":SOUND 10,10
6070 E=0:RETURN
6075 REM
6500 REM
6505 REM *TARGET ROUTINE"
6510 REM
6515 FOR X=1 TO 8
6520 FOR Y=1 TO 8
6525 PRINT@480,"X";X;" Y";Y;" -";
6530 INPUT A$:SOUND 210,2
6535 IF A$="" THEN GOTO 6550
6540 IF A$="Q" THEN E=0:RETURN
6545 CW$(X,Y)=A$
6550 NEXT Y
6555 NEXT X
6560 PRINT@480,"[TARGET FINISHED]":SOUND 10,10
6565 E=0:RETURN
7000 REM
7005 REM *COORDINATE CONVERTER ROUTINE*
7010 REM
7015 NX=INSTR(N+1,C$,S$)
7020 IF NX=0 THEN E=2:RETURN
7025 X$=MID$(C$,N+1,1)
7030 NY=INSTR(NX+1,C$,S$)
7035 IF NY=0 THEN E=2:RETURN
7040 Y$=MID$(C$,NX+1,1)
7045 REM *CONVERT VALUES*
7050 IF X$="1" THEN X=0:GOTO 7095
7055 IF X$="2" THEN X=32:GOTO 7095
7060 IF X$="3" THEN X=64:GOTO 7095
7065 IF X$="4" THEN X=96:GOTO 7095
7070 IF X$="5" THEN X=128:GOTO 7095
7075 IF X$="6" THEN X=160:GOTO 7095
7080 IF X$="7" THEN X=192:GOTO 7095
7085 IF X$="8" THEN X=224:GOTO 7095
7090 E=2:RETURN
7095 IF Y$="1" THEN Y=0:GOTO 7140
7100 IF Y$="2" THEN Y=32:GOTO 7140
7105 IF Y$="3" THEN Y=64:GOTO 7140
7110 IF Y$="4" THEN Y=96:GOTO 7140
7115 IF Y$="5" THEN Y=128:GOTO 7140
7120 IF Y$="6" THEN Y=160:GOTO 7140
7125 IF Y4="7" THEN Y=192:GOTO 7140
7130 IF Y$="8" THEN Y=224:GOTO 7140
7135 E=2:RETURN
7140 E=0:RETURN
7500 REM
7505 REM *MOTION ROUTINE*
7510 REM
7515 IF DRX=0 AND GT=1 THEN POKE 49154,
     7:GOTO 7525
7520 IF DRX=1 AND GT=1 THEN POKE 49154,0
7525 FOR C=1 TO TPX
7530 POKE 49152,7:POKE 49152,6:POKE 49152,7
7535 NEXT C
7540 IF DRX=0 THEN CPX=CPX-TPX:GOTO 7550
7545 CPX=CPX+TPX
7550 IF DRY=0 AND GT=1 THEN POKE 49154,
     7:GOTO 7560
7555 IF DRY=1 AND GT=1 THEN POKE 49154,0
7560 FOR C=1 TO TPY
7565 POKE 49152,7:POKE 49152,5:POKE 49152,7
7570 NEXT C
7575 IF DRY=0 THEN CPY=CPY-TPY:GOTO 7585
7580 CPY=CPY+TPY
7585 IF GT=1 THEN GOTO 7640
7590 IF PT=0 THEN RETURN
7595 POKE 49154,0
7600 FOR C=1 TO 32
7605 POKE 49152,7:POKE 49152,3:POKE 49152,7
7610 NEXT C
7615 POKE 49154,68
7620 FOR C=1 TO 32
7625 POKE 49152,7:POKE 49152,3:POKE 49152,7
7630 NEXT C
7635 RETURN
7640 POKE 49154,64
7645 FOR C=1 TO 32
7650 POKE 49152,7:POKE 49152,3:POKE 49152,7
7655 NEXT C
7660 POKE 49154,4
7665 FOR C=1 TO 32
7670 POKE 49152,7:POKE 49152,3:POKE 49152,7
7675 NEXT C
7680 RETURN
```

Let's get into the description of how the simulator works. Fig. 4-29 is a representation of what the graphics screen shows on the TI system. Located in the center of the screen is the 8 × 8 workspace area. Above this area is a representation of a gantry arm very much like the actual XY table. On power-up and initial run, this arm is shown with its Y and X axes in the X-1,Y-1 or HOME position. At the bottom of the screen is a light-colored text window. This window will allow six lines of text to be displayed simultaneously. It is through this window that operator commands and program prompt and error messages are shown. As the commands are implemented the graphic representation of the gantry arm is actually moved on the screen to other locations. Fig. 4-30 is a close-up of the 8 × 8 block representation, showing the arm in a different position after a MOVE command. Black and white blocks are shown graphically, superimposed on the

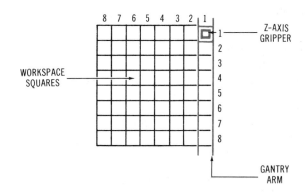

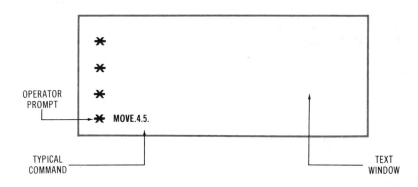

Fig. 4-29. Representation of screen of XY table simulator.

8 × 8 array. The simulator has the ability to physically pick up blocks and move them to other spaces on the screen through the use of special graphics MOVE instructions.

Let's get into the initialization portion of the program. Fig. 4-31 is the portion of the listing covering initialization. In actuality, this program has two initialization routines. This one refers to the variable default conditions and general initialization. A second one refers to graphics initialization. In the figure, line 20 clears the screen. Notice in this BASIC, that being Texas Instruments BASIC, a CALL CLEAR instruction will do the job of clearing the screen whereas in MicroSoft BASIC the CLS does the same thing. The next instruction, DISPLAY AT, will print on the screen, at location row ten, column eight, the message "One moment please". This is a prompting message to an operator that has to wait for the initialization to go through. In this version of BASIC it takes a good 20 seconds before the initialization routine has completed. Line 35 calls a machine-language routine called INIT, which will initialize the Texas Instrument's memory system to receive a machine-language BASIC linkable program. The following line, 40, actually loads in the machine-language program from disk. The program is called SCROLL. We'll be going into the detailed discussion of this program later on. From line 55 on, various variables that will be used throughout the program are set to their initial values. The first three being DIMension statements much like in the coordinate controller program for

current world, target world, and program line. You'll notice that in this system, current world and target world are not string variables. They are simply numeric variables with eight X and eight Y locations for a total of 64. Line 70 sets up current X position, current Y position. These are pixel locations on the graphic screen. Where the initial X,Y arm is located. Lines 144 and 41 define the HOME position pixel. We'll get into this in a lot more detail later, also. Various other variables are set on this line, as you can see. Line 75 is a portion of the error-handling routines from the coordinate

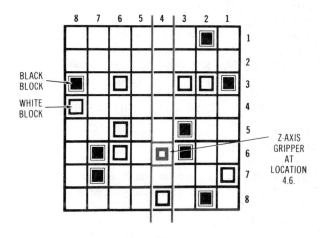

Fig. 4-30. Close-up of 8 × 8 workspace showing gantry arm.

```
20 CALL CLEAR
25 DISPLAY AT(10,8)SIZE(17):"ONE MOMENT PLEASE"
30 REM
35 CALL INIT
40 CALL LOAD("DSK1.SCROLL")
45 REM
50 REM
55 REM ***VARIABLE DEFAULTS***
60 REM
65 DIM CW(8,8):: DIM TW(8,8):: DIM PL$(52)
70 CX=144 :: CY=41 :: C=0 :: SW=0 :: CR=1 :: E=0 :: PC=0 :: G=0
75 S$="." :: EN$="[INVALID COMMAND]" :: BN$="[BAD INPUT VALUE]"
```

Fig. 4-31. Listing of initialization routine.

controller program. The error-handling prompts or messages shown on the screen are enacted in this program as string variables: EN$ being invalid command and BN$ being bad input value. You will see how these are used throughout the program.

So, between line 20 and line 80, this is the general initialization routine. During this time the computer will be loading the machine-language SCROLL routine and setting up variable defaults. From here the program will go on to initialize the graphics system.

It is here that some explanation of the Texas Instruments particular graphics system should be given. It will help in customizing the system to your particular hardware requirements. The basic TI99/4A graphics display consisted of 256 pixels horizontally by 192 pixels vertically. Pixels are small addressable dots on the screen. The TI system allowed 16 colors for each of these small entities. In the TI system, characters are designed as eight horizontal pixels wide by eight vertical pixels high, for a total of 64 pixels. Each character can be defined by the user to represent any symbol or shape. Upon power-up the Texas Instruments system assigns a number of these characters to standard ASCII shapes. During the initialization phase of the graphics program, extra characters are defined. These characters are used to shape the graphics screen to show the 8 × 8 or 64 workspace squares, to show the gantry XY arm and to build the six-line text window. I will explain more as we get into the graphics initialization routine. This portion of the program can be found in Fig. 4-32.

In the first line of the subprogram, line 95 is the first you see of the call character statement. Realizing that the Texas Instruments machine that I am describing has been discontinued, I will explain fully the actions of each unusual BASIC statement, so that you will be able to adapt it to a particular system of your own. The character subprogram allows you to define these special graphics characters. You can redefine the standard set of characters and undefined characters. The standard codes are from 32 to 95. You will notice that in this line we are redefining character code number 91. The numbers within the parentheses, after the character code, represent bits of the character. The first two numbers (in this particular one they are all zeros) define the first line of eight pixels, the next two the second line, and so on down to the final two to be the bottom eight pixels. The character itself is set up, and the way the numbers within the parentheses are defined is shown in Fig. 4-33. This figure shows a typical character block with the rows one through eight and the locations within the quotes for each value. There are four left blocks and four right blocks. Each set of four is represented as a hexadecimal character within the quotes. Looking down to line 100, character code number 96, all Fs, represents all pixels on. Looking at the diagram of the character block, you will see that an FF in row one will define two 4-bit blocks of ones. Moving down to line 105, character code 97, in the second set of numbers, 81, will define row two as having the leftmost block on and the rightmost block of the right blocks on. Go through the numbers and see if you can correlate the shapes of each character.

Let's go down the list and define what each character represents within the program. Line 95, the character code number 91, is a space code, no pixels on for the text window. This will be defined with a special color later. The character code 96 in lne 100 is used as a fill-in block on the gripper small square within the X gantry arm. You'll see that in much greater detail later. In line 105, character code 97, is used to define the workspace squares (the 8 × 8 blocks in the center of the screen). Line 110, character code 104, is used to define the gripper. Line 120, character code 112, as you can see, is a very long 64-character entry character definition. This refers to something called a *sprite*. A sprite is a 16 × 16 pixel block. It is four normal character blocks laid side by side, two on top of two. A sprite can be moved around the screen pixel by pixel in any direction. This is how we get smooth movement of the gantry arm. Character code 112 represents the vertical bars of this gantry arm piece. As we get into the painting of the gantry on the screen, I will explain how the sprite maps onto the areas of the screen. Let's move along to line 125, character code 120 is a special null background character that is painted across the entire screen on power-up. You will see this background character happening during the One Moment Please prompting message. Line 130 through 165, character codes 120 through 128 define black on a light-blue background characters, which are the numbers one through eight, which label the workspace squares on the outside edges. Line 170, character code 136, is a representation of a black block, the type of block that might be placed on a workspace square. Conversely, line 175, character code

```
80 REM
85 REM ***CHARACTER DEFINITIONS***
90 REM
95 CALL CHAR(91,"0000000000000000")
100 CALL CHAR(96, "FFFFFFFFFFFFFFFF")
105 CALL CHAR(97, "FF818181818181FF")
110 CALL CHAR(104, "7050507000000000")
115 CALL CHAR(106, "FF000000FF000000000000000000000
    00FF000000FF00000000000000000000")
120 CALL CHAR(112, "8888888888888888888888888888888
    000000000000000000000000000000000")
125 CALL CHAR(120, "0000000000000000")
130 CALL CHAR(121, "2060202020207000")
135 CALL CHAR(122, "7080803040080F800")
140 CALL CHAR(123, "F80810300888870000")
145 CALL CHAR(124, "1030509F8101000")
150 CALL CHAR(125, "F880F008088870000")
155 CALL CHAR(126, "384080F088887000")
160 CALL CHAR(127, "F8081020404040000")
165 CALL CHAR(128, "7088887088887000")
170 CALL CHAR(136, "007E7E7E7E7E7E00")
175 CALL CHAR(137, "FF818181818181FF")
180 CALL CHAR(138, "0020200000000000")
185 CALL CHAR(139, "0070707000000000")
```

Fig. 4-32. Listing of graphics initialization routine.

Lines 190 through 230, as shown in Fig. 4-34, are CALL COLOR statements. In TI BASIC these statements will specify a foreground and background color for certain sets of characters. Looking at line 205, the character set represented there is set 14. Character set 14 corresponds to 136 through

```
190 REM
195 REM ***COLOR SET DEFINITIONS***
200 REM
205 CALL COLOR(14,2,16)
210 CALL COLOR(0,2,15,1,2,15,2,2,15,3,2,15,4,2,15,
    5,2,15,6,2,15,7,2,15,8,2,15)
215 CALL COLOR(9,13,12)
220 CALL COLOR(10,7,12)
225 CALL COLOR(12,2,1,13,2,1)
230 REM
```

Fig. 4-34. Listing of color definition routine.

143; therefore, the white block, the black block, the cursor for the PLOT, and the fill-in for the white block are all represented by this line. The foreground color specified is a two that corresponds to the color black. The background color, 16, corresponds to the color white. A list of color codes and what they represent is shown in Table 4-5. You

Table 4-5. TI/994A Color Set Definition

COLOR	CODE	COLOR	CODE
TRANSPARENT	1	MEDIUM RED	9
BLACK	2	LIGHT RED	10
MEDIUM GREEN	3	DARK YELLOW	11
LIGHT GREEN	4	LIGHT YELLOW	12
DARK BLUE	5	DARK GREEN	13
LIGHT BLUE	6	MAGENTA	14
DARK RED	7	GRAY	15
CYAN	8	WHITE	16

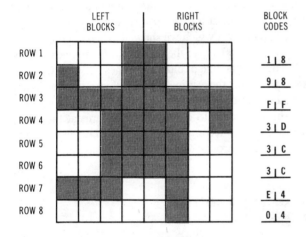

CALL CHAR (96, "1898FF3D3C3CE404")

Fig. 4-33. Representation of CALL CHAR operation.

137, is the white block representation. Line 180, code 138, is a small vertical cursor that is just a few pixels wide by a few pixels tall, which is the center of one of the workspace blocks. This little cursor is used when the operator is entering block assignments in the current world and target world routines. Line 185, character code 139, represents the fill-in portion to the middle of the gripper Z-axis square on the screen, for a white block. When a white block is being carried, the middle of this gripper must be filled in to show that it is carrying this block. When it is holding a white block, this character code is specified. Moving along, the next portion of the program is where you define the colors each one of these characters will carry when represented on the screen.

can refer back to this throughout the program to help you rewrite it for you own particular BASIC. Also, in Table 4-6 is a list of character sets and their corresponding character codes. The next line, 210, specifies quite a lot of character sets. Each one of these sets, however, uses the same fore-

Table 4-6. TI/994A Character Set Definition

SET	ASCII CODE	SET	ASCII CODE
0	30-31	8	88-95
1	32-39	9	96-103
2	40-47	10	104-111
3	48-55	11	112-119
4	56-63	12	120-127
5	64-71	13	128-135
6	72-79	14	136-143
7	80-87		

ground and background color scheme. Foreground being black the background being gray. If you will notice, from the line, the first number in a set of three will be the character set number. So, sets 0, 1, 2, 3, 4, 5, 6, 7, and 8 will all be represented by black and gray. Character sets 0 through 8 represent the standard ASCII characters that are normally produced by the Texas Instruments machine. A portion of set 8, however, has been defined, in line 95, as the text window space code 91. As you will notice from the table on character sets, set 8 will handle ASCII codes through number 95; therefore, the text window showing all zeros will show all background that is gray. Line 215 sets the color set for character set 9. Character set 9 will include the CALL CHAR statement, code 96 in line 100, code 97 in line 105, and that about does it. The first one is the fill-in for the gripper during the carry of the black block. The second one is the representation of the workspace squares. Looking at the color scheme for this, we have 13 being the foreground color, which is dark green. The background color is light yellow. The next color set to be defined is in line 220, which is set number 10, which will include character codes 104-111. That would be the gripper. This color scheme shows a foreground of dark red and a background of light yellow. Finally, line 225 contains the character sets 12 and 13, which bring us down through the numbers 1 through 8 that are located around the workspace squares. These colors are black and transparent. Transparent, on the Texas Instruments screen, looks like a sky blue. Now we've finished defining the various characters for use on the screen and their color sets; the next thing to look at is the actual running of the program.

After beginning to run a BASIC program, there are typically many messages on the screen. The first job of the program is to clear the screen. Because we are using various color tones and shapes, this particular program will shade the screen with what looks like a clear or transparent color but in actuality it is our special character for background. Fig. 4-35 is the short subprogram, from lines 35 to 0, that will place the

```
235 REM
240 REM ***BLANK SCREEN WITH SPECIAL CHARACTER***
245 REM
250 FOR Y=1 TO 24
255 CALL HCHAR(Y,1,120,32)
260 NEXT Y
```

Fig. 4-35. Listing of blank-screen routine.

special character, code 120, in every location of the screen. It does this through a FOR-NEXT loop, which will count 24 rows and will place a character 32 times across on columns. This represents the 24 × 32 arrangement of the TI screen.

The CALL H character is the horizontal character, the Y location is the row specified in the FOR-NEXT loop, the next location is the X value, the next is the character code, and the next is how many times to repeat it.

After the screen has been cleared with the special character, the 8 × 8 workspace grid will be drawn onto the screen. Lines 270 through 345 are the draw blocks grid routine. It

```
265 REM
270 REM ***DRAW BLOCKS GRID***
275 REM
280 FOR Y=6 TO 13
285 CALL HCHAR(Y,12,97,8)
290 NEXT Y
295 X=12
300 FOR C=128 TO 121 STEP -1
305 CALL HCHAR(4,X,C)
310 X=X+1
315 NEXT C
320 Y=6
325 FOR C=121 TO 128
330 CALL VCHAR(Y,21,C)
335 Y=Y+1
340 NEXT C
```

Fig. 4-36. Listing of draw-blocks-grid routine.

can be seen in Fig. 4-36. The first thing that is done is the blocks themselves are drawn onto the screen. That is done from lines 280 to 290 much like the screen was cleared with a special character; the H character command is used to replicate blocks eight at a time horizontally. The block here is character code 97, which we showed called out on line 105.

From lines 300 to 315 the numbers eight to one are drawn on top of this first line of blocks. In lines 320 to 340, the end of the program, the vertical line of blocks on the right are labeled one through eight in ascending order. From here the next thing to do is to draw the gantry arm at the HOME position. This is accomplished with the draw gantry routine from lines 350 through 400. This is shown in Fig. 4-37. Here

```
350 REM
355 REM **DRAW GANTRY**
360 REM
365 REM
370 CALL MAGNIFY(2)
375 N=1
380 FOR Y=41 TO 96 STEP 16
385 CALL SPRITE(#N,112,16,Y,144)
390 N=N+1
395 NEXT Y
400 CALL SPRITE(#9,104,7,41,144)
```

Fig. 4-37. Listing of draw-gantry routine.

we come to a representation on the screen that will take some explanation.

The Texas Instruments graphics system allows things called sprites. As discussed before sprites are 16 × 16 pixel blocks that can move, pixel by pixel, very smoothly across the screen. Sprites are much larger than regular characters and can be magnified. Therefore, I have used sprites to draw the vertical arms of the gantry. As you can see in line 370, the CALL MAGNIFY command will magnify, times 2, the sprite that has been created in line 120, character code 112.

The normal 16×16 will then take up the space of the 32×32. This little routine, draw gantry, basically draws the two vertical lines on the screen in the sprite mode so that they can be moved very smoothly.

Finally, the gray text window, where all the commands are entered, is drawn on the screen with the routine in lines 410 through 435 (shown in Fig. 4-38). This is the draw text window routine. It also uses the H character command doing

```
410 REM
415 REM ***DRAW TEXT WINDOW***
420 REM
425 FOR Y=17 TO 22
430 CALL HCHAR(Y,6,91,20)
435 NEXT Y
```

Fig. 4-38. Listing of draw-text-window routine.

the text window special-character-number 91. It'll draw a block, as you can see in the previous figures, to where text is entered. From here on out the program will look very similar to the previous coordinate controller program. The text input handler, from lines 460 to 500, shown in Fig. 4-39, is using the same design philosophy; however, TI BASIC statements are used. Notice in line 475, the display prompt asterisk is shown at the bottom left of the text window, which is location row 21, column 6. The size statement there says that it is only one character long. DISPLAY@ is like a PRINT@ command. Line 480, ACCEPT@, allows the program to show the blinking cursor at the place right next to the asterisk. VALIDATE will only allow upper alpha, which is the UALPHA command, and NUMERIC commands. In MicroSoft BASIC that's not necessary. Size, being 16, will allow 16 entries. That is also not necessary in MicroSoft BASIC. It will beep much like the SOUND command that we used before, and as you can see, the command will be stored in variable C$. From there line 485 will catapult to a subroutine located at line 505. This first subroutine is a SCROLL routine that we will go over later. Suffice to say, at this time, that when this subroutine is entered, the text will scroll up one line and then at line 490, will check to see if nothing has been entered, in which case it will, once again, prompt with an asterisk. Line 495, with a GOSUB 810, will jump the program to the command interpreter section. After a valid command has been entered and after the command has been executed, the program will jump back to line 475, which will prompt with another asterisk.

From here, let's look at the SCROLL routine. Fig. 4-40 is an assembly listing written in TMS9900 Microprocessor Assembly Language. It constitutes the SCROLL routine. The purpose of this routine is to scroll four lines of command language within the text window. The reason a machine-language routine was used here is because BASIC, in that it is an interpretive language, would be too slow, and the operator would notice each character being moved. Within the TI99/4A there is a block of memory that is used to store

```
460 REM
465 REM ***TEXT INPUT HANDLER***
470 REM
475 DISPLAY AT(21,6)SIZE(1):"*"
480 ACCEPT AT(21,7)VALIDATE(UALPHA,NUMERIC)SIZE(16)
    BEEP:C$
485 GOSUB 505
490 IF C$="" THEN GOTO 475
495 GOSUB 810
500 GOTO 475
```

Fig. 4-39. Listing of text input-handler routine.

the screen contents. What the SCROLL routine does is move, byte by byte, the contents of the screen up one line location. There are four lines of text that must be moved, as I said before. The top line is called line 4. As you can see, in line 14 of the assembly listing, the data in memory that represents line 3 is pulled and then written over that of line 4, shown in line 16. Then the data, from the memory locations representing line 2, are written over the previous data from line 3. The same happens from line 1 to line 2. That leaves you with line 1 and line 2 showing the same information. Line 26 of the assembly listing then builds a line of blank characters. This blank line is then moved, as shown in line 28 of the listing, to the current lines data place, which is line 1. At the bottom it then returns to BASIC. This is an assembly-language routine that is called from BASIC using the LINK BASIC statement. Each time the SCROLL routine is called, it will move line 3 to 4, 2 to 3, 1 to 2, and then blank line 1; move back to the BASIC listing that will then write a new asterisk for a prompt. You will notice, throughout the BASIC listing, this subroutine will be called every time input is requested.

From here let's look at line 495. The GOSUB 810, as I said before, will move the program down to the command interpreter. At this time, let's go to line 810. This subroutine is shown in Fig. 4-41. The command interpreter is the routine that takes the ASCII string entity in the command string, or C$, that was entered in by the operator when the prompt asterisk was shown, and then looks through the string for the first period delimiter. This routine is very much like the one in the coordinate controller written in MicroSoft BASIC.

Looking at the routine, line 820, the position, or POS, statement will place the numbered position within the string that the period delimiter, which is located in S$, is found. From here, line 825 checks to see if any period delimiter was found. If none were to be found, N would be zero and, therefore, you would get an error condition of two (if you remember correctly, that is a bad input value) and then it would go to an error handling routine. Line 830 is where it is determined that the number is not zero, there is indeed a period. That particular segment o C$, from position 1 to the position before the period, or N − 1, is then moved into W$. From here, we go into another routine, which is still shown on this figure. This routine is called check entry against table. It is almost exactly the same routine as shown

```
99/4 ASSEMBLER
VERSION 1.2
0001
0002                          * SCROLL ROUTINE
0003                          *
0004                          * TO SCROLL 4 LINES OF COMMAND LANGUAGE
0005                          *
0006                                   DEF SCROLL
0007                          *
0008          2024            VMBW    EQU     >2024
0009          202C            VMBR    EQU     >202C
0010          837C            STATUS  EQU     >837C
0011                          *
0012   0000   0200            SCROLL  LI      R0, >245    Load up parameters
       0002   0245
0013   0004   0201                    LI      R1, >2800
       0006   2800
0014   0008   0202                    LI      R2, >14     Get line 3 data
       000A   0014
0015   000C   0420                    BLWP    @VMBR
       000E   202C
0016   0010   0200                    LI      R0, >225    Write it over that of line 4
       0012   0225
0017   0014   0420                    BLWP    @VMBW
       0016   2024
0018   0018   0200                    LI      R0, >265    Get line 2 data
       001A   0265
0019   001C   0420                    BLWP    @VMBR
       001E   202C
0020   0020   0200                    LI      R0, >245    Write it over that of line 3
       0022   0245
0021   0024   0420                    BLWP    @VMBW
       0026   2024
0022   0028   0200                    LI      R0, >285    Get line 1 data
       002A   0285
0023   002C   0420                    BLWP    @VMBR
       002E   202C
0024   0030   0200                    LI      R0, >265    Write it over that of line 2
       0032   0265
0025   0034   0420                    BLWP    @VMBW
       0036   2024
0026   0038   0200                    LI      R0, >2A5    Get the blank line data
       003A   02A5
0027   003C   0420                    BLWP    @VMBR
       003E   202C
0028   0040   0200                    LI      R0, >285    Erase the current line's data
       0042   0285
0029   0044   0420                    BLWP    @VMBW
       0046   2024
0030   0048   04C0                    CLR     R0          Prepare to return to BASIC
0031   004A   D800                    MOVB    R0, @STATUS
       004C   837C
0032   004E   045B                    RT
0033                                  END
  R0           0000      R1       0001      R10      000A      R11      000B
  R12          000C      R13      000D      R14      000E      R15      000F
  R2           0002      R3       0003      R4       0004      R5       0005
  R6           0006      R7       0007      R8       0008      R9       0009
D SCROLL       0000      STATUS   837C      VMBR     202C      VMBW     2024
0000 ERRORS
```

Fig. 4-40. Assembly-language listing of SCROLL routine.

```
810 REM **COMMAND INTERPRETER**
815 REM
820 N=POS(C$,S$,1)
825 IF N=0 THEN E=2 :: GOTO 910
830 W$=SEG$(C$,1,N-1)
835 REM
840 REM *CHECK ENTRY AGAINST TABLE*
845 REM
850 IF W$="EDIT" THEN GOSUB 1005
855 IF W$="EXECUTE" THEN GOSUB 1085
860 IF W$="GOTO" THEN GOSUB 1090
865 IF W$="GET" THEN GOSUB 1140
870 IF W$="HOME" THEN GOSUB 1200
875 IF W$="LIST" THEN GOSUB 1225
880 IF W$="MOVE" THEN GOSUB 1400
885 IF W$="PLACE" THEN GOSUB 1525
890 IF W$="PLOT" THEN GOSUB 1590
895 IF W$="PROGRAM" THEN GOSUB 1705
900 IF W$="RUN" THEN GOSUB 1795
905 IF W$="TARGET" THEN GOSUB 1875
910 IF E=1 THEN DISPLAY AT(21,6)SIZE(17):BN$ ::
    CALL SOUND(250,200,5) :: GOSUB 505
915 IF E=2 THEN DISPLAY AT(21,6)SIZE(17):EN$ ::
    CALL SOUND(250,200,5) :: GOSUB 505
920 E=0 :: RETURN
```

Fig. 4-41. Listing of command-interpreter routine.

```
1700 REM
1705 REM **PROGRAM INPUT ROUTINE**
1710 REM
1715 DISPLAY AT(21,6)SIZE(12):"NEW PROGRAM?"
1720 ACCEPT AT(21,19)VALIDATE("YN")SIZE(1)BEEP:PI$
1725 GOSUB 505
1730 IF PI$="N" THEN GOTO 1785
1735 FOR PC=1 TO 50
1740 DISPLAY AT(21,6)SIZE(3):" > ";PC
1745 ACCEPT AT(21,10)VALIDATE(UALPHA,NUMERIC)SIZE
     (11)BEEP:PI$
1750 GOSUB 505
1755 IF PI$="Q" THEN RETURN
1760 IF PI$="" THEN GOTO 1740
1765 PL$(PC)=PI$
1770 NEXT PC
1775 DISPLAY AT(21,6)SIZE(8):"MEM FULL" :: CALL
     SOUND(250,200,5):: GOSUB 505
1780 RETURN
1785 GOTO 1005
```

Fig. 4-42. Listing of program-input routine.

```
1000 REM
1005 REM **EDIT PROGRAM ROUTINE**
1010 REM
1015 DISPLAY AT(21,6)SIZE(6):"LINE#?"
1020 ACCEPT AT(21,13)VALIDATE(NUMERIC, "Q")SIZE(2)
     BEEP:PC$ :: GOSUB 505
1025 ON ERROR 1020
1030 IF PC$="Q" THEN RETURN
1035 PC=VAL(PC$)
1040 EL$=PL$(PC)
1045 DISPLAY AT(21,6)SIZE(16):EL$ :: GOSUB 505
1050 DISPLAY AT(21,6)SIZE(3):" > ";PC
1055 ACCEPT AT(21,10)VALIDATE(UALPHA,NUMERIC)SIZE
     (11)BEEP:EL$
1060 GOSUB 505
1065 IF EL$="Q" THEN RETURN
1070 PL$(PC)=EL$
1075 GOTO 1015
1080 REM
```

Fig. 4-43. Listing of EDIT routine.

before, except for the GOSUB locations. As you can see, lines 850 through line 905, W$ is checked for valid commands from edit through target. Lines 910 and 915 are a portion of the error-handling routines for this listing. You'll notice that in line 910, if E = 1, which is an invalid command, the display will be BN$ and if you'll remember, up at the initialization routine, BN$ was bad input value. If E = 2, it will display EN$, which is an invalid command. Notice the CALL SOUND routine that will basically do the same thing as the SOUND routine in MicroSoft BASIC.

Assuming that you have a valid command, when that command is noticed, somewhere between line 850 and line 905, the program will jump to the subroutine that handles that command. At this time, let's go through them and look at each listing individually. In keeping with the last discussion, let's go through the listing the same way as in the coordinate controller program. Therefore, the first routine that we should look at is the program input routine, which goes to the subroutine at line 1705. This routine can be seen in Fig. 4-42. Looking at the figure, we see that the first action the program does is to display a prompt asking if this is a new program. Second line, 1720, will then look for a yes or a no supplied as a Y or an N. At this point, automatically, the subroutine at line 505 will be called, remember this is the SCROLL routine. From here, the input that the operator typed, which is now in PI$, or program input string, is checked for a no conition. If it is a no condition then that means it is not a new program; it is an existing program. From that point, it will be sent down to line 1785, which is another GOTO that is to line 1005, which is the edit routine. Assuming that a yes was input, line 1735 will then initiate a program counter from line 1 through 50. You'll recognize this from the program input routine of the coordinate controller. At line 1740 the greater than sign (>) prompt is then input onto the screen. Beside it, the initial line number, which will be line 1 and then line 1745, will be looking for the actual command string and put it into PI$. Line 1750 will scroll up a line. Lines 1755 and 1760 will check for either a quit or a null input, and the actual storage line of 1765 will put away a valid command string into the program line array, which is PL$ in a memory location pointed to by the program counter. From there, line 1770 will increment the program counter, and the whole routine will start again. As you can see, this is almost exactly the same as the coordinate controller program except for the fact that it is written in TI BASIC.

The next routine to go over is the EDIT routine. Remembering that if a no condition was entered during the prompt go to the edit routine. Of course, it is possible to enter the edit routine at any time from the command prompt. The edit routine is shown in Fig. 4-43. The initial line of 1015 displays the prompt asking for the line number you'd like to edit. From here, line 1020 will then look for a valid line number. Notice at the end of line 1020, there is another statement stuck on. That is the SCROLL routine. Line 1030 checks to see if your input was a Q or a quit condition. From here, the operator input is turned into an actual value with the VALUE statement of line 1035. Line 1040 will pull the program line, at program counter location, out of the program line array in memory and put it into a variable called EL$, which stands for edit line. Line 1045 will then display that line and scroll. This gives the operator the ability to see what is in that location before changing it. Line 1050 then gives a prompt, with the current program counter value, which will equal the same value as the line above this prompt. Line 1055 then asks the operator to type in the new contents of that program counter line. Line 1060 scrolls again. Line 1065 then looks for a quit condition. If no quit is entered, then in line 1070 the newly edited line will be put back into the program line array at the program counter location. Line 1075 starts the routine all over again.

From here, the next routine we should cover is the LIST routine. Fig. 4-44 shows the contents of the LIST routine. Each one of these routines is designed, specifically, to act like the routines in the coordinate controller. Once again, in line 1235 the LIST routine will ask you if you want this particular listing printed. Line 1240 looks for a Y or an N, standing for yes or no. If the answer is yes, the program will then jump to the print routine, which we will go over in a minute. Assuming that you want the screen to show the listing, a no would be the answer, and line 1255 would be the next program line executed. In this line the program counter is set up to look at lines 1 through 51. Remember from the discussion before, that the number 51 was chosen so that you could show three lines at a time (fifty-one being divisible by three equally). The FOR-TO loop in 1255 and 1260, which displays the current program counter value and then increments them one at a time, is the same mechanism as shown in the coordinate controller listings previously. This goes on through to line 1275. Line 1280 stops the program and asks if you want three more lines displayed. At this point nothing but a beep will show up for the operator. You have the possibility of typing a Q at this time, which will throw you out of the program. If the Q is not entered, the next three lines

```
1220 REM
1225 REM **LIST PROGRAM ROUTINE**
1230 REM
1235 DISPLAY AT(21,6)SIZE(6):"PRINT?"
1240 ACCEPT AT(21,13)VALIDATE("YN")SIZE(1)BEEP:A$
1245 GOSUB 505
1250 IF A$="Y" THEN GOTO 1305
1255 FOR PC=1 TO 51 STEP 3
1260 FOR FC=1 TO 3
1265 DISPLAY AT(21,6)SIZE(16):PC+(FC-1);" ";PL$(PC+(FC-1))
1270 GOSUB 505
1275 NEXT FC
1280 ACCEPT AT(21,6)VALIDATE(" Q")SIZE(1)BEEP:A$
1285 GOSUB 505
1290 IF A$="Q" THEN RETURN
1295 NEXT PC
1300 RETURN
1305 DISPLAY AT(21,6)SIZE(14):"PRINTER READY?"
1310 ACCEPT AT(21,21)VALIDATE("YNQ")SIZE(1)BEEP:A$
1315 GOSUB 505
1320 IF A$="Q" THEN RETURN
1325 IF A$="N" THEN GOTO 1305
1330 OPEN #1: "RS232.BA=9600.PA=N.DA=7.TW.LF"
1335 DISPLAY AT(21,6)SIZE(13):"PROGRAM NAME:"
1340 GOSUB 505
1345 ACCEPT AT(21,6)VALIDATE(UALPHA,NUMERIC)SIZE(16)BEEP:A$
1350 GOSUB 505
1355 PRINT #1:A$
1360 PRINT #1
1365 FOR PC=1 TO 50
1370 PRINT #1:PC;" ";PL$(PC)
1375 IF PL$(PC)= "END." THEN GOTO 1385
1380 NEXT PC
1385 CLOSE #1
1390 RETURN
```

Fig. 4-44. Listing of LIST routine.

will appear and this will go on all the way up to line 51.

The next part of the routine, starting at line 1305, is the printer output subroutine. You will notice that in line 1305, the prompt asking if the printer is ready is displayed. Line 1310 looks for a Y, an N, or a Q for quit. If quit is entered, of course, the subroutine will end. If a no is entered, it will ask you again, PRINTER READY? And it will keep on asking until the printer is ready or you quit. If the printer is ready, line 1330 is specific to the TI system. It opens a file called number 1. It is RS232 with a baud rate 9600, no parity, data bit 7, two stop bits and a line feed. Your system may require a whole different set of prerequisite setups. Line 1335, after the printer is ready, will prompt the operator for a program name to be printed on the paper before the listing is printed. After that is entered, in line 1345 the routine will then begin to print. The first thing it prints, in line 1355, is the program name that you just entered. Line 1360 will carriage return the output that will give a space between the listing and the name. Line 1365 sets up the program counter again. Line 1370 will print the data in the first location pointed at the program counter at the program line array. Line 1375 will check to see if that was an END statement. In this command language END would be the end of the program. In the RUN routine, you will see that, when the word END happens, the program will stop. In this case, the word END is encountered, the listing will stop. If the command is not an END, 1380 will increment the program counter. This will keep going from lines 1 to 50. After line 50, 1385 will close out the file. Line 1390 will then return to the calling program.

Speaking of the RUN routine, the next listing, shown in Fig. 4-45, is the RUN subprogram. Looking at the figure we see

```
1790 REM
1795 REM **RUN PROGRAM ROUTINE**
1800 REM
1805 FOR PC=1 TO 50
1810 CALL KEY(0,K,S)
1815 IF S=1 THEN RETURN
1820 C$=PL$(PC)
1825 DISPLAY AT(21,6)SIZE(16):PC;" ";C$
1830 GOSUB 505
1835 IF C$="END." OR C$="" THEN GOTO 1855
1840 GOSUB 810
1845 IF G=1 THEN G=0 :: GOTO 1820
1850 NEXT PC
1855 DISPLAY AT(21,6)SIZE(12):"PROGRAM STOP"
1860 GOSUB 505
1865 RETURN
1870 REM
```

Fig. 4-45. Listing of RUN routine.

that, once again, in line 1805, the program counter is set up. Line 1810, the CALL KEY instruction is an immediate look at the keyboard. If you'll recall back to the coordinate controller in Fig. 4-17, that RUN routine had an INKEY$ statement that was an immediate look at the keyboard. Similarly, this routine, with the CALL KEY, looks for a keypress. If a keypress is found, then the S variable in that statement will be a one. If this is true, the RUN routine will stop. This is the run abort mechanism. If, in line 1810 and 1815, a key is not found to be depressed, then line 1820 will pull the first program line in the program line array and put it into command string (C$). C$ is the variable that the command interpreter works on. In line 1825, what was just pulled from the array is displayed on the screen so that the operator can get a listing of what is going on at the present time. Line 1835 checks to see if this is an END or a nothing. If it is an END or a nothing, it will stop the program. As you can see in line 1855, where it will jump to, the words "PROGRAM STOP" will be displayed on the screen, the screen will be scrolled, and the routine will return. If it is not an END or a null input, line 1840 tells the routine to go to the command interpreter subroutine, which is at line 810. When it comes back from the command interpreter, line 1845 will check to see if a GOTO indication happened during that command. If it did, it will then clear out the GOTO (G) flag. From here, it goes to the next program counter location. Of course, if it is a GOTO routine, it will then take the program counter value from that routine. This will go along until the end of the program counter lines and then return.

```
1090 REM **GOTO ROUTINE**
1095 REM
1100 GX=POS(C$,S$,N+1)
1105 IF GX=0 THEN E=1 :: RETURN
1110 GN$=SEG$(C$,N+1,2)
1115 GN=VAL(GN$)
1120 PC=GN
1125 G=1
1130 RETURN
```

Fig. 4-46. Listing of GOTO routine.

Fig. 4-46 shows the GOTO subroutine. As I said previously, if a GOTO condition exists, the G flag will be raised, saying that the previous command has the next program counter value. Let's see how the TI BASIC subroutine handles this. It starts off with line 1100, pulling the actual line number from the command string. If you remember, the last time we looked at the command string or when we found out that it was a GOTO routine, the variable N represented the numerical position, within the string, of the first period delimiter. From here, we would like to look at the locations 1 + the variable N to the next period delimiter which would, in this case, be the end of the line number input by the operator. Line 1100 does this by finding the position of the delimiter in S$, which is, as we know, the period. In line C$, from the number in variable N + 1, this new position is shown to be put into variable GX. Line 1105 checks to see that GX is not a zero, meaning there is no line number shown after the GOTO, which would be an error condition, which is E = 1 and, as I said before, one means a bad input value. If there is a line number, then line 1110 will take that segment of C$, from the right of the first period, two value locations. These two value locations will show a number 1 through 50. Line 1115 then makes this string value into an actual numeric

```
1135 REM
1140 REM **GET BLOCK ROUTINE**
1145 REM
1150 GOSUB 1400
1155 C=VAL(X$):: R=VAL(Y$)
1160 BC=CW(R,C)
1165 IF BC=137 THEN CR=16
1170 IF BC=136 THEN CR=2
1175 CALL SPRITE(#10,139,CR,Y,X)
1180 CALL HCHAR(R+5,20-C,97)
1185 CW(R,C)=97
1190 RETURN
```

Fig. 4-47. Listing of GET BLOCK routine.

```
1395 REM
1400 REM **MOVE ROUTINE**
1405 REM
1410 GOSUB 1995 :: REM CALL GET X,Y
1415 IF E=1 THEN RETURN
1420 REM
1425 REM *MAKE THE MOVE*
1430 REM
1435 IF X=CX THEN XV=0
1440 IF X < CX THEN XV=-1
1445 IF X > CX THEN XV=1
1450 IF Y=CY THEN YV=0
1455 IF Y < CY THEN YV=-1
1460 IF Y > CY THEN YV=1
1465 GOSUB 2155 :: REM SET IN MOTION
1470 IF XV=0 THEN GOTO 1490
1475 CALL COINC($4,89,X,1,XC)
1480 IF XC=-1 THEN XV=0 :: GOSUB 2155
1485 GOTO 1495
1490 IF YV=0 THEN GOTO 1515
1495 CALL POSITION(#9,PY,PX)
1500 IF PY <> Y THEN GOTO 1470
1505 YV=0 :: GOSUB 2155
1510 GOTO 1470
1515 CX=X :: CY=Y :: RETURN
```

Fig. 4-48. Listing of MOVE routine.

value. Line 1120 replaces the program counter with this numeric value, raises the G flag and returns. From here, we get to our first significantly different subroutine. The GET subroutine in Fig. 4-47 manipulates graphics on screen instead of motors on the XY table. Let's look at the structure of this routine.

The GET BLOCK routine, shown here, will only move the gantry and Z position gripper on the screen. What you don't see in this listing is the subroutine called out at line 1150 that says GOSUB 1400. Line 1400 is the motion routine. I'll go over that later, but it should suffice to say, right now, that the X and Y positions are pulled off from the GET routine. The gantry arm is then moved. From here, it jumps back to this routine where, in line 1155, the values X$ and Y$ are put into a column and row variables of C and R, represented at that point by current world (CW). If you remember from the last time we went over the coordinate controller, the operator will input where white and black blocks are in the routine by sticking them into a current world array, which is arranged 8 × 8 to represent the workspace squares. Here, the row-and-column counters are pointed to a certain location in the current world. The number that is now sitting at this location is then entered into BC. From here, lines 1165 and 1170 check to see if it is a white or a black block. A white block would be a current world value of 137, and a black block would be 136. At this point, either 1165 or 1170 determines the variable CR or the color of the block to be shown within the little window of the gripper. Line 1175 will then move the gripper to that XY location. Line 1180 will then change that particular work-space square to an empty value that will be a light green in this case. It will then change the current world location to a value of 97, which is an empty. Realizing that this whole routine is simulating the gripper traveling down in a vertical direction, grabbing a block and bringing it up, which would leave that current location empty.

From here, let's go look at the motion routine. It is shown in Fig. 4-48. Realize, of course, that the motion or MOVE routine is in itself a command. MOVE will only move the gantry to a particular XY location. This routine is really made up of two routines; the second part is shown in Fig. 4-49, which we'll get to in a minute. Looking at Fig. 4-48, the MOVE routine starts off jumping to a subroutine at location 1995. This is the portion of the routine that we will get to later. Basically, that portion gets the X and Y values. Line 1415 will check to see if an error condition existed in those X and Y values. Notice, that the error being one would be a bad input value and that the error prompt would then be displayed on the monitor. Starting at line 1425 the actual move happens. Line 1435 through line 1460 check to see if the current values of X and Y are equal to, greater than, or less than the requested change in X or Y. If the current X location is less than the requested X location, then as you can see, the XV or direction variable will then equal a positive value or a one. The opposite is true if the current X location is greater than the requested location. Then a −1 would be the direction. The same is true for the Y inputs. Line 1465 goes to another subroutine that actually sets the arm in motion on the screen. This is also shown in Fig. 4-49. We will get to this later, also. Line 1470 checks to see if the velocity direction is zero. In this case it will jump down to see if the Y velocity is zero, and then nothing will happen. This is the case that, if the operator accidentally tells the program to move to a location that it is already at, it is to ignore the command. Assuming that this is not the case, line 1475 realizes that the previous line (1465) set the arm in motion. Here in line 1475, the CALL COINCIDENCE statement is an on-going check that is like a routine in itself that will check the screen contents, and when the gantry arm has reached the current X location, it will then give a certain indication in variable XC. If XC is a −1, meaning that there is a coincidence on the screen between the requested X value and the new gantry position, the X velocity will then turn to zero. It sends that to the motion subroutine in line 2155. This will stop the gantry from traveling in the X direction. Line 1495 will check to see which position the Y portion of the gantry is. If it is not the

```
1990 REM
1995 REM *GET X,Y VALUES*
2000 REM
2005 NX=POS(C$,S$,N+1)
2010 IF NX=0 THEN E=1 :: RETURN
2015 X$=SEG$(C$,N+1,1)
2020 NY=POS(C$,S$,NX+1)
2025 IF NY=0 THEN E=1 :: RETURN
2030 Y$=SEG$(C$,NX+1,1)
2035 REM
2040 REM *CONVERT VALUES*
2045 REM
2050 IF X$="1" THEN X=144 :: GOTO 2095
2055 IF X$="2" THEN X=136 :: GOTO 2095
2060 IF X$="3" THEN X=128 :: GOTO 2095
2065 IF X$="4" THEN X=120 :: GOTO 2095
2070 IF X$="5" THEN X=112 :: GOTO 2095
2075 IF X$="6" THEN X=104 :: GOTO 2095
2080 IF X$="7" THEN X=96 :: GOTO 2095
2085 IF X$="8" THEN X=88 :: GOTO 2095
2090 E=1 :: RETURN
2095 IF Y$="1" THEN Y=41 :: GOTO 2140
2100 IF Y$="2" THEN Y=49 :: GOTO 2140
2105 IF Y$="3" THEN Y=57 :: GOTO 2140
2110 IF Y$="4" THEN Y=65 :: GOTO 2140
2115 IF Y$="5" THEN Y=73 :: GOTO 2140
2120 IF Y$="6" THEN Y=81 :: GOTO 2140
2125 IF Y$="7" THEN Y=89 :: GOTO 2140
2130 IF Y$="8" THEN Y=97 :: GOTO 2140
2135 E=1 :: RETURN
2140 E=0 :: RETURN
2145 REM
2150 REM **MOTION ROUTINE**
2155 REM
2160 CALL MOTION(#1,0,XV,#2,0,XV,#3,0,XV,#4,0,XV,#9,YV,XV,#10,YV,XV)
2165 RETURN
2170 REM
```

Fig. 4-49. Listing of MOVE routine (part 2).

requested Y location, it will go on. When it does reach the requested position, it makes the Y velocity equal zero, which will stop the Y direction. From here, the requested X and Y locations now become the current X and Y positions in line 1515 and returns. Now to fill in this void, let's look at Fig. 4-49. The first thing we see is, in line 1995, the GET X and Y values routine. This takes the command string, finds the X and Y values in between the period delimiters, and puts them into X$ and Y$. This happens between lines 2005 through 2030. Now, these are numeric values from 1 to 8. The screen values, however, are pixel values from 88 to 144. Each one of these locations is the upper leftmost dot of the 8 × 8 block. The Y locations go from number 41 through 97. Therefore, this routine, from line 2040 through line 2145, will convert the operators one through eight into actual screen locations for the graphic routine. From here the motion routine is automatically entered (you can see it in line 2150).

Motion, in line 2160 (the CALL MOTION statement in TI BASIC), will allow sprites to move, pixel by pixel, on the screen. There are four sprites that make up the long parallel gantry arm. If you will notice, after the opening parenthesis, the #1 indicates sprite number 1, which is the uppermost parallel-bars portion of the gantry. The next number represents the row velocity or the top-to-bottom velocity. In the case of the parallel-bar gantry, it only goes from left to right or right to left, it does not go up and down; therefore, this velocity would be set to zero. I am calling that the Y direction. The next, XV, is the X velocity. It is now how fast you want it to move, in which direction, plus or minus, and, as you can see, these same three variables go on throughout the parentheses in line 2160 until there are two more sprites that are identified, numbers 9 and 10. Sprite number 9 is the actual block in the middle that represents the gripper and Z axis. This moves along with the X and Y axes. Sprite number 10 would be any block that is being carried. As you can see, this motion routine sets in motion and then returns. It is the MOVE routine that determines when it should start and we went over that previously.

From here, let's get on to the PLACE routine. As I said before, the PLACE routine is very similar to the GET routine. Shown in Fig. 4-50, PLACE starts business in line 1535. Here it once again asks to go to the subroutine at 1400, which will get the X and Y values. Everything else looks very familiar until you get to line 1550. This line checks to see if that particular location is full (there is already a block in it). If this is the case, a bad input value prompt will be

```
1520 REM
1525 REM **PLACE BLOCK ROUTINE**
1530 REM
1535 GOSUB 1400
1540 C=VAL(X$):: R=VAL(Y$)
1545 BS=CW(R,C)
1550 IF BS=137 OR BS=136 THEN E=1 :: RETURN
1555 IF CR=16 THEN BK=137
1560 IF CR=2 THEN BK=136
1565 CALL HCHAR(R+5,20-C,BK)
1570 CALL SPRITE(#10,96,1,Y,X):: CR=1
1575 CW(R,C)=BK
1580 RETURN
```

Fig. 4-50. Listing of PLACE BLOCK routine.

displayed on the screen. Assuming that the workspace area is empty, then line 1555 will check to see if the color block that it is holding is white or black. At that point it will put either the white or the black block in that current workspace location on the video screen. It will then, in line 1575, replace the current world row-and-column value with this new block color and then return.

From here, let's get the HOME routine out of the way before plotting the current and target worlds. The HOME routine is shown in Fig. 4-51. As you can see, this is a very

```
1195 REM
1200 REM **HOME ROUTINE**
1205 REM
1210 X=144 :: Y=41
1215 GOTO 1425
```

Fig. 4-51. Listing of HOME routine.

simple routine. It starts off by placing the X and Y values at the upper right workspace square locations on the video screen (pixel location 144 and 41 of the X and Y row and column locations on the screen). It then goes to line 1425, which is the portion of the MOVE routine that actually moves the gantry arm. As you know, from the previous work on the coordinate converter, the HOME routine will take the gantry arm from wherever it is currently and move it to the upper right-hand work-space square valued at X position 1, Y position 1.

Now, let's get into the PLOT and TARGET routines. The PLOT routine, which allows the operator to enter white and black blocks into the current world, is shown in Fig. 4-52. Looking at the figure, we see something called CALL DELSPRITE in line 1600. What this does is get rid of all the sprites on the screen at this point. To date the only sprites shown would be the gantry, the arm, and any block that it might be carrying. Then in line 1605, those familiar values of X, 144 and Y, 41, are set up. Row and column numbers are set up to be one, and a value called SW is equal to zero. Line 1610 sends out a new sprite, numbered 6. This sprite is a small, operator-moved cursor. Line 1615 then goes to a subroutine shown at line 670. We have not gone over this subroutine as yet. It is shown in Fig. 4-53. Basically, what it does (if you'll go to that figure now) is it will, on the fly, look

```
1585 REM
1590 REM **PLOT ROUTINE**
1595 REM
1600 CALL DELSPRITE(ALL)
1605 X=144 :: Y=41 :: R=1 :: C=1 :: SW=0
1610 CALL SPRITE(#6,138,13,Y,X)
1615 GOSUB 670
1620 IF CM=81 THEN GOTO 1660
1625 IF CM=87 THEN BK=137 :: GOTO 1645
1630 IF CM=66 THEN BK=136 :: GOTO 1645
1635 IF CM=32 THEN BK=97 :: GOTO 1645
1640 CALL SOUND(250,200,5):: GOTO 1615
1645 CALL HCHAR(R+5,20-C,BK)
1650 CW(R,C)=BK
1655 GOTO 1615
1660 N=1
1665 FOR Y1=41 TO 96 STEP 16
1670 CALL SPRITE(#N,112,16,Y1,CX)
1675 N=N+1
1680 NEXT Y1
1685 CALL SPRITE(#9,104,7,CY,CX)
1690 CALL DELSPRITE(#6)
1695 RETURN
```

Fig. 4-52. Listing of PLOT routine.

```
665 REM
670 REM **CURSOR ROUTINE(PLOT COMMANDS)**
675 REM
680 CALL KEY(0,CM,S)
685 IF S=0 THEN GOTO 680
690 IF CM>11 THEN RETURN
695 IF CM=11 THEN R=R-1 :: IF R<1 THEN R=1
700 IF CM=8 THEN C=C+1 :: IF C>8 THEN C=8
705 IF CM=10 THEN R=R+1 :: IF R>8 THEN R=8
710 IF CM=9 THEN C=C-1 :: IF C<1 THEN C=1
715 IF SW=1 THEN GOTO 755
720 X$=STR$(C):: Y$=STR$(R):: GOSUB 2040
725 XP=X :: YP=Y
730 CALL SPRITE(#6,138,13,YP,XP)
735 GOTO 680
```

Fig. 4-53. Listing of CURSOR routine.

```
1870 REM
1875 REM **TARGET PLOT ROUTINE**
1880 REM
1885 GOSUB 930
1890 X=232 :: Y=41 :: SW=1 :: R=1 :: C=1
1895 CALL SPRITE(#6,138,13,Y,X)
1900 GOSUB 670
1905 IF CM=81 THEN GOTO 1945
1910 IF CM=87 THEN BK=137 :: GOTO 1930
1915 IF CM=66 THEN BK=136 :: GOTO 1930
1920 IF CM=32 THEN BK=97 :: GOTO 1930
1925 CALL SOUND(250,200,5)
1930 CALL HCHAR(R+5,31-C,BK)
1935 TW(R,C)=BK
1940 GOTO 1900
1945 CALL DELSPRITE(#6)
1950 FOR Y=6 TO 13
1955 CALL HCHAR(Y,23,120,8)
1960 NEXT Y
1965 RETURN
1970 REM
1975 DISPLAY AT(21,6)SIZE(17):I$
1980 GOSUB 505
1985 GOTO 475
```

Fig. 4-54. Listing of TARGET plot routine.

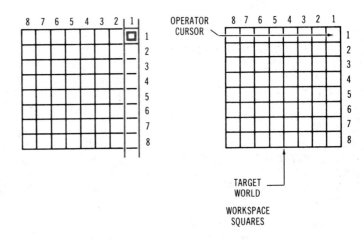

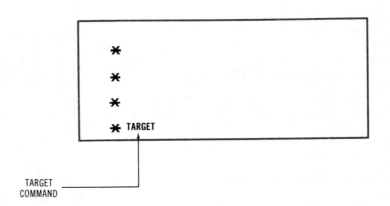

Fig. 4-55. Representation of screen showing target world.

for a keypress at the keyboard. What it is looking for is the arrow cursor position keys. When it finds either an up, down, right, or left cursor key, it will move sprite 6 around the workspace squares, depending on where the operator lets go of the key. It also looks for the keys B or W. If it finds B, it will assume that the operator wishes to place a black block in that particular square. The same is true for a W for a white block. It also looks for the value Q. If Q is entered, it will then quit the PLOT routine and go back to the command handler. All this happens from line 680 through line 710.

Going back to the PLOT routine in Fig. 4-52, you can see that the actual motion of the cursor sprite happens here. In the end, when a Q is detected, line 1670 and line 1685 will then restore the gantry and the Z axis arm. Line 1690 will then get rid of the current cursor sprite.

The TARGET PLOT routine works in a very similar manner. Not to belabor the logic involved, I have just decided to show this routine in its entirety in Fig. 4-54. Similarly, the TARGET routine calls another routine at line 930, which will display an alternate workspace to the right of the current world workspace. A representation of the screen at this point is shown in Fig. 4-55, and this routine is shown in Fig. 4-56. This alternate space represents the target world workspace. When the Q is entered, the target world workspace will then be erased from the screen and the prompt command handler will then be entered.

For clarity the entire listing of the XY table simulator is shown in Fig. 4-57.

As you have seen in this chapter, building an XY table, controlling, and simulating it can be very stimulating. The next two chapters will focus on available hobbyist robot systems for the home. These systems can be used to train would-be robotics experts in various control techniques. We will look at the different systems, home-in on the HERO1 robot from Heathkit, and give you a feel for the final area of advanced robot systems.

```
925 REM
930 REM **DISPLAY TARGET WORLD ROUTINE**
935 REM
940 X=23 :: Y=6
945 FOR R=1 TO 8
950 FOR C=8 TO 1 STEP -1
955 BK=TW(R,C)
960 IF BK=0 THEN BK=97
965 CALL HCHAR(Y,X,BK)
970 X=X+1 :: NEXT C
975 X=23 :: Y=Y+1 :: NEXT R
980 RETURN
```

Fig. 4-56. Listing of TARGET routine (part 2).

```
10 REM
11 REM
12 REM ****XY TABLE SIMULATOR****
13 REM
14 REM c 1983 M. J. ROBILLARD
15 REM
16 REM
20 CALL CLEAR
25 DISPLAY AT(10,8)SIZE(17):"ONE MOMENT PLEASE"
30 REM
35 CALL INIT
40 CALL LOAD("DSK1.SCROLL")
45 REM
50 REM
55 REM ***VARIABLE DEFAULTS***
60 REM
65 DIM CW(8,8):: DIM TW(8,8):: DIM PL$(52)
70 CX=144 :: CY=41 :: C=0 :: SW=0 :: CR=1 ::
   E=0 :: PC=0 :: G=0
75 S$="." :: EN$="[INVALID COMMAND]" :: BN$=
   "[BAD INPUT VALUE]"
80 REM
85 REM ***CHARACTER DEFINITIONS***
90 REM
95 CALL CHAR(91,"00000000000000000")
100 CALL CHAR(96,"FFFFFFFFFFFFFFFF")
105 CALL CHAR(97,"FF818181818181FF")
110 CALL CHAR(104,"7050507000000000")
115 CALL CHAR(106,"FF000000FF000000000000000000000
    000FF000000FF0000000000000000000000")
120 CALL CHAR(112,"8888888888888888888888888888
    8880000000000000000000000000000")
125 CALL CHAR(120,"0000000000000000")
130 CALL CHAR(121,"2060202020207000")
135 CALL CHAR(122,"708808304080F800")
140 CALL CHAR(123,"F808103008887000")
145 CALL CHAR(124,"10305090F8101000")
150 CALL CHAR(125,"F880F00808887000")
155 CALL CHAR(126,"384080F088887000")
160 CALL CHAR(127,"F808102040404000")
165 CALL CHAR(128,"7088887088887000")
170 CALL CHAR(136,"007E7E7E7E7E7E00")
175 CALL CHAR(137,"FF818181818181FF")
180 CALL CHAR(138,"0020200000000000")
185 CALL CHAR(139,"0070707000000000")
190 REM
195 REM ***COLOR SET DEFINITIONS***
200 REM
205 CALL COLOR(14,2,16)
210 CALL COLOR(0,2,15,1,2,15,2,2,15,3,2,15,4,2,
    15,5,2,15,6,2,15,7,2,15,8,2,15
215 CALL COLOR(9,13,12)
220 CALL COLOR(10,7,12)
225 CALL COLOR(12,2,1,13,2,1)
230 REM
235 REM
240 REM **BLANK SCREEN WITH SPECIAL CHARACTER**
245 REM
250 Y=1 TO 24
255 CALL HCHAR(Y,1,120,32)
260 NEXT Y
265 REM
270 REM   ***DRAW BLOCKS GRID***
275 REM
280 FOR Y=6 TO 13
285 CALL HCHAR(Y,12,97,8)
290 NEXT Y
295 X=12
300 FOR C=128 TO 121 STEP -1
305 CALL HCHAR(4,X,C)
310 X=X+1
315 NEXT C
320 Y=6
325 FOR C=121 TO 128
330 CALL VCHAR(Y,21,C)
335 Y=Y+1
340 NEXT C
345 REM
350 REM
355 REM **DRAW GANTRY**
360 REM
365 REM
370 CALL MAGNIFY(2)
375 N=1
380 FOR Y=41 TO 96 STEP 16
385 CALL SPRITE(#N,112,16,Y,144)
390 N=N+1
395 NEXT Y
400 CALL SPRITE(#9,104,7,41,144)
405 REM
```

Fig. 4-57. Entire listing of XY table simulator.

Fig. 4-57. Cont.

```
410 REM
415 REM ***DRAW TEXT WINDOW***
420 REM
425 FOR Y=17 TO 22
430 CALL HCHAR(Y,6,91,20)
435 NEXT Y
440 REM
445 REM
450 REM
455 REM
460 REM
465 REM ***TEXT INPUT HANDLER***
470 REM
475 DISPLAY AT(21,6)SIZE(1):"*"
480 ACCEPT AT(21,7)VALIDATE(UALPHA,NUMERIC)SIZE
    (16)BEEP:C$
485 GOSUB 505
490 IF C$="" THEN GOTO 475
495 GOSUB 810
500 GOTO 475
505 REM ***SUBROUTINE SECTION***
510 REM
515 REM **SCROLL TEXT AREA**
520 REM
525 CALL LINK("SCROLL")
530 RETURN
535 REM
540 REM
545 REM **CONVERT ROW-COLUMN(CURRENT WORLD)**
550 REM
555 REM *GET X VALUE*
560 REM
565 IF PX=1 THEN X=19
570 IF PX=2 THEN X=18
575 IF PX=3 THEN X=17
580 IF PX=4 THEN X=16
585 IF PX=5 THEN X=15
590 IF PX=6 THEN X=14
595 IF PX=7 THEN X=13
600 IF PX=8 THEN X=12
605 REM
610 REM *GET Y VALUE*
615 REM
620 IF PY=1 THEN Y=6
625 IF PY=2 THEN Y=7
630 IF PY=3 THEN Y=8
635 IF PY=4 THEN Y=9
640 IF PY=5 THEN Y=10
645 IF PY=6 THEN Y=11
650 IF PY=7 THEN Y=12
655 IF PY=8 THEN Y=13
660 RETURN
665 REM
670 REM **CURSOR ROUTINE(PLOT COMMANDS)**
675 REM
680 CALL KEY(0,CM,S)
685 IF S=0 THEN GOTO 680
690 IF CM>11 THEN RETURN
695 IF CM=11 THEN R=R-1 :: IF R<1 THEN R=1
700 IF CM=8 THEN C=C+1 :: IF C>8 THEN C=8
705 IF CM=10 THEN R=R+1 :: IF R>8 THEN R=8
710 IF CM=9 THEN C=C-1 :: IF C<1 THEN C=1
715 IF SW=1 THEN GOTO 755
720 X$=STR$(C):: Y$=STR$(R):: GOSUB 2040
725 XP=X :: YP=Y
730 CALL SPRITE(#6,138,13,YP,XP)
735 GOTO 680
740 REM
745 REM **CONVERT ROW-COLUMN(TARGET WORLD)**
750 REM
755 IF C=1 THEN X=232
760 IF C=2 THEN X=224
765 IF C=3 THEN X=216
770 IF C=4 THEN X=208
775 IF C=5 THEN X=200
780 IF C=6 THEN X=192
785 IF C=7 THEN X=184
790 IF C=8 THEN X=176
795 Y$=STR$(R):: GOSUB 2095
800 GOTO 725
805 REM
810 REM **COMMAND INTERPRETER**
815 REM
820 N=POS(C$,S$,1)
825 IF N=0 THEN E=2 :: GOTO 910
830 W$=SEG$(C$,1,N-1)
835 REM
840 REM *CHECK ENTRY AGAINST TABLE*
845 REM
850 IF W$="EDIT" THEN GOSUB 1005
```

Fig. 4-57. Cont.

```
855  IF W$="EXECUTE" THEN GOSUB 1085
860  IF W$="GOTO" THEN GOSUB 1090
865  IF W$="GET" THEN GOSUB 1140
870  IF W$="HOME" THEN GOSUB 1200
875  IF W$="LIST" THEN GOSUB 1225
880  IF W$="MOVE" THEN GOSUB 1400
885  IF W$="PLACE" THEN GOSUB 1525
890  IF W$="PLOT" THEN GOSUB 1590
895  IF W$="PROGRAM" THEN GOSUB 1705
900  IF W$="RUN" THEN GOSUB 1795
905  IF W$="TARGET" THEN GOSUB 1875
910  IF E=1 THEN DISPLAY AT(21,6)SIZE(17):BN$ ::
     CALL SOUND(250,200,5):: GOSUB 505
915  IF E=2 THEN DISPLAY AT(21,6)SIZE(17):EN$ ::
     CALL SOUND(250,200,5):: GOSUB 505
920  E=0 :: RETURN
925  REM
930  REM **DISPLAY TARGET WORLD ROUTINE**
940  X=23 :: Y=6
945  FOR R=1 TO 8
950  FOR C=8 TO 1 STEP -1
955  BK=TW(R,C)
960  IF BK=0 THEN BK=97
965  CALL HCHAR(Y,X,BK)
970  X=X+1 :: NEXT C
975  X=23 :: Y=Y+1 :: NEXT R
980  RETURN
985  REM
990  REM
995  REM **COMMAND HANDLERS**
1000 REM
1005 REM **EDIT PROGRAM ROUTINE**
1010 REM
1015 DISPLAY AT(21,6)SIZE(6):"LINE#?"
1020 ACCEPT AT(21,13)VALIDATE(NUMERIC,"Q")SIZE
     (2)BEEP:PC$ :: GOSUB 505
1025 ON ERROR 1020
1030 IF PC$="Q" THEN RETURN
1035 PC=VAL(PC$)
1040 EL$=PL$(PC)
1045 DISPLAY AT(21,6)SIZE(16):EL$ :: GOSUB 505
1050 DISPLAY AT(21,6)SIZE(3):" > ";PC
1055 ACCEPT AT(21,10)VALIDATE(UALPHA,NUMERIC)
     SIZE(11)BEEP:EL$
1060 GOSUB 505
1065 IF EL$="Q" THEN RETURN
1070 PL$(PC)=EL$
1075 GOTO 1015
1080 REM
1085 GOTO 1975
1090 REM **GOTO ROUTINE**
1095 REM
1100 GX=POS(C$,S$,N+1)
1105 IF GX=0 THEN E=1 :: RETURN
1110 GN$=SEG$(C$,N+1,2)
1115 GN=VAL(GN$)
1120 PC=GN
1125 G=1
1130 RETURN
1135 REM
1140 REM **GET BLOCK ROUTINE*
1145 REM
1150 GOSUB 1400
1155 C=VAL(X$):: R=VAL(Y$)
1160 BC=CW(R,C)
1165 IF BC=137 THEN CR=16
1170 IF BC=136 THEN CR=2
1175 CALL SPRITE(#10,139,CR,Y,X)
1180 CALL HCHAR(R+5,20-C,97)
1185 CW(R,C)=97
1190 RETURN
1195 REM
1200 REM **HOME ROUTINE**
1205 REM
1210 X=144 :: Y=41
1215 GOTO 1425
1220 REM
1225 REM **LIST PROGRAM ROUTINE**
1230 REM
1235 DISPLAY AT(21,6)SIZE(6):"PRINT?"
1240 ACCEPT AT(21,13),VALIDATE("YN")SIZE(1)BEEP:
     A$
1245 GOSUB 505
1250 IF A$="Y" THEN GOTO 1305
1255 FOR PC=1 TO 51 STEP 3
1260 FOR FC=1 TO 3
1265 DISPLAY AT(21,6)SIZE(16):PC+(FC-1);" ";PL$
     (PC+(FC-1))
1270 GOSUB 505
1275 NEXT FC
```

Fig. 4-57. Cont.

```
1280 ACCEPT AT(21,6)VALIDATE(" Q")SIZE(1)BEEP:A$
1285 GOSUB 505
1290 IF A$="Q" THEN RETURN
1295 NEXT PC
1300 RETURN
1305 DISPLAY AT(21,6)SIZE(14):"PRINTER READY?"
1310 ACCEPT AT(21,21)VALIDATE("YNQ")SIZE(1)BEEP:
     A$
1315 GOSUB 505
1320 IF A$="Q" THEN RETURN
1325 IF A$="N" THEN GOTO 1305
1330 OPEN #1:"RS232.BA=9600.PA=N.DA=7.TW.LF"
1335 DISPLAY AT(21,6)SIZE(13):"PROGRAM NAME:"
1340 GOSUB 505
1345 ACCEPT AT(21,6)VALIDATE(UALPHA,NUMERIC)SIZE
     (16)BEEP:A$
1350 GOSUB 505
1355 PRINT #1:A$
1360 PRINT #1
1365 FOR PC=1 TO 50
1370 PRINT #1:PC;" ";PL$(PC)
1375 IF PL$(PC)="END." THEN GOTO 1385
1380 NEXT PC
1385 CLOSE #1
1390 RETURN
1395 REM
1400 REM **MOVE ROUTINE**
1405 REM
1410 GOSUB 1995 :: REM CALL GET X,Y
1415 IF E=1 THEN RETURN
1420 REM
1425 REM *MAKE THE MOVE*
1430 REM
1435 IF X=CX THEN XV=0
1440 IF X < CX THEN XV=-1
1445 IF X > CX THEN XV=1
1450 IF Y=CY THEN YV=0
1455 IF Y < CY THEN YV=-1
1460 IF Y > CY THEN YV=1
1465 GOSUB 2155 :: REM SET IN MOTION
1470 IF XV=0 THEN GOTO 1490
1475 CALL COINC(#4,89,X,1,XC)
1480 IF XC=-1 THEN XV=0 :: GOSUB 2155
1485 GOTO 1495
1490 IF YV=0 THEN GOTO 1515
1495 CALL POSITION(#9,PY,PX)
1500 IF PY <> Y THEN GOTO 1470
1505 YV=0 :: GOSUB 2155
1510 GOTO 1470
1515 CX=X :: CY=Y :: RETURN
1520 REM
1525 REM **PLACE BLOCK ROUTINE**
1530 REM
1535 GOSUB 1400
1540 C=VAL(X$):: R=VAL(Y$)
1545 BS=CW(R,C)
1550 IF BS=137 OR BS=136 THEN E=1 :: RETURN
1555 IF CR=16 THEN BK=137
1560 IF CR=2 THEN BK=136
1565 CALL HCHAR(R+5,20-C,BK)
1570 CALL SPRITE(#10,96,1,Y,X):: CR=1
1575 CW(R,C)=BK
1580 RETURN
1585 REM
1590 REM **PLOT ROUTINE**
1595 REM
1600 CALL DELSPRITE(ALL)
1605 X=144 :: Y=41 :: R=1 :: C=1 :: SW=0
1610 CALL SPRITE(#6,138,13,Y,X)
1615 GOSUB 670
1620 IF CM=81 THEN GOTO 1660
1625 IF CM=87 THEN BK=137 :: GOTO 1645
1630 IF CM=66 THEN BK=136 :: GOTO 1645
1635 IF CM=32 THEN BK=97 :: GOTO 1645
1640 CALL SOUND(250,200,5):: GOTO 1615
1645 CALL HCHAR(R+5,20-C,BK)
1650 CW(R,C)=BK
1655 GOTO 1615
1660 N=1
1665 FOR Y1=41 TO 96 STEP 16
1670 CALL SPRITE(#N,112,16,Y1,CX)
1675 N=N+1
1680 NEXT Y1
1685 CALL SPRITE(#9,104,7,CY,CX)
1690 CALL DELSPRITE(#6)
1695 RETURN
1700 REM
1705 REM **PROGRAM INPUT ROUTINE**
1710 REM
1715 DISPLAY AT(21,6)SIZE(12):"NEW PROGRAM?"
```

Fig. 4-57. Cont.

```
1720 ACCEPT AT(21,19)VALIDATE("YN")SIZE(1)BEEP:
     PI$
1725 GOSUB 505
1730 IF PI$="N" THEN GOTO 1785
1735 FOR PC=1 TO 50
1740 DISPLAY AT(21,6)SIZE(3):" > ";PC
1745 ACCEPT AT(21,10)VALIDATE(UALPHA,NUMERIC)
     SIZE(11)BEEP:PI$
1750 GOSUB 505
1755 IF PI$="Q" THEN RETURN
1760 IF PI$="" THEN GOTO 1740
1765 PL$(PC)=PI$
1770 NEXT PC
1775 DISPLAY AT(21,6)SIZE(8):"MEM FULL" :: CALL
     SOUND(250,200,5):: GOSUB 505
1780 RETURN
1785 GOTO 1005
1790 REM
1795 REM **RUN PROGRAM ROUTINE**
1800 REM
1805 FOR PC=1 TO 50
1810 CALL KEY(0,K,S)
1815 IF S=1 THEN RETURN
1820 C$=PL$(PC)
1825 DISPLAY AT(21,6)SIZE(16):PC;" ";C$
1830 GOSUB 505
1835 IF C$="END." OR C$="" THEN GOTO 1855
1840 GOSUB 810
1845 IF G=1 THEN G=0 :: GOTO 1820
1850 NEXT PC
1855 DISPLAY AT(21,6)SIZE(12):"PROGRAM STOP"
1860 GOSUB 505
1865 RETURN
1870 REM
1875 REM **TARGET PLOT ROUTINE**
1880 REM
1885 GOSUB 930
1890 X=232 :: Y=41 :: SW=1 :: R=1 :: C=1
1895 CALL SPRITE(#6,138,13,Y,X)
1900 GOSUB 670
1905 IF CM=81 THEN GOTO 1945
1910 IF CM=87 THEN BK=137 :: GOTO 1930
1915 IF CM=66 THEN BK=136 :: GOTO 1930
1920 IF CM=32 THEN BK=97 :: GOTO 1930
1925 CALL SOUND (250,200,5)
1930 CALL HCHAR(R+5,31-C,BK)
1935 TW(R,C)=BK
1940 GOTO 1900
1945 CALL DELSPRITE(#6)
1950 FOR Y=6 TO 13
1955 CALL HCHAR(Y,23,120,8)
1960 NEXT Y
1965 RETURN
1970 REM
1975 DISPLAY AT(21,6)SIZE(17):I$
1980 GOSUB 505
1985 GOTO 475
1990 REM
1995 REM *GET X,Y VALUES*
2000 REM
2005 NX=POS(C$,S$,N+1)
2010 IF NX=0 THEN E=1 :: RETURN
2015 X$=SEG$(C$,N+1,1)
2020 NY=POS(C$,S$,NX+1)
2025 IF NY=0 THEN E=1 :: RETURN
2030 Y$=SEG$(C$,NX+1,1)
2035 REM
2040 REM *CONVERT VALUES*
2045 REM
2050 IF X$="1" THEN X=144 :: GOTO 2095
2055 IF X$="2" THEN X=136 :: GOTO 2095
2060 IF X$="3" THEN X=128 :: GOTO 2095
2065 IF X$="4" THEN X=120 :: GOTO 2095
2070 IF X$="5" THEN X=112 :: GOTO 2095
2075 IF X$="6" THEN X=104 :: GOTO 2095
2080 IF X$="7" THEN X=96 :: GOTO 2095
2085 IF X$="8" THEN X=88 :: GOTO 2095
2090 E=1 :: RETURN
2095 IF Y$="1" THEN Y=41 :: GOTO 2140
2100 IF Y$="2" THEN Y=49 :: GOTO 2140
2105 IF Y$="3" THEN Y=57 :: GOTO 2140
2110 IF Y$="4" THEN Y=65 :: GOTO 2140
2115 IF Y$="5" THEN Y=73 :: GOTO 2140
2120 IF Y$="6" THEN Y=81 :: GOTO 2140
2125 IF Y$="7" THEN Y=89 :: GOTO 2140
2130 IF Y$="8" THEN Y=97 :: GOTO 2140
2135 E=1 :: RETURN
2140 E=0 :: RETURN
```

Fig. 4-57. Cont.

```
2145 REM
2150 REM **MOTION ROUTINE**
2155 REM
2160 CALL MOTION(#1,0,XV,#2,0,XV,#3,0,XV,#4,0,
     XV,#9,YV,XV,#10,YV,XV)
2165 RETURN
2170 REM
```

Chapter 5

Personal Robot Hardware

Throughout the last four chapters I have been presenting design philosophy and practices for a number of robotic applications. We have ventured from the AGV that scurries about the factory floor delivering parts and retrieving finished goods, discussed security robots and their special functions, presented operations programs for automated mail carriers that beep at the office door, and finally, covered the XY table and the programs that allowed it to rearrange a world of blocks. In all these discussions the emphasis was on industrial or commercial uses for robots. As you know, science-fiction writers have for years dramatized a different sort of robot. This species usually appeared somewhat humanlike, and it also possessed humanlike capabilities of mobility, communication, and in some cases, artificial human emotion. Today robots emulating these qualities have shown up, mostly, in a new segment of the market.

PERSONAL ROBOTICS

What makes a robot personal? Has anyone asked this question in regard to personal computers? Certainly human qualities are not a prerequisite. Any electronic equipment designed for general consumer use is usually deemed personal. You would think that a robot designed for use in the home would be called a home robot. The term *personal robotics* has already been used so frequently that it would be pointless to change the direction now.

What is the difference between a *personal* robot and an *industrial* robot? The answer to this question really depends on the type of robot. There are small manipulators that have been called personal robot equipment that can be applied to several light-duty industrial jobs. The key here is the words light-duty. Generally, the cost of the robot is directly proportional to the load-carrying capacity and/or its accuracy. Obviously, the cost of the robot will categorize it more fully. Therefore, low-cost, light-duty, and somewhat accurate robots might be classified as personal robots.

Is this description correct or can there be other reasons for personal items? Obviously, anything used in the home or on a highly personal basis could be classified personal. In the case of robots, engineers using a small, light-duty, industrial robot at their desks or in an office environment might use the term "personal robot." In this chapter *personal robots* refers to those designed for use in the home. Today several manufacturers are suppling robots fully assembled and in kit form for use by experimenters, students, and general hobbyists. Their prices range from somewhere around $250 all the way up to the $10,000 range. With such a varied cost range, you would think that "personal" would refer only to those in the under $1000 market, however, today's trend toward multithousand dollar personal computers should dispel any problems with accepting a $5000 personal robot. Where did these personal robots come from and where are they going? Let's take a brief journey through the history of personal robots.

Ever since experimentation in electronics and electromechanics has been underway, hobbyists have been building personal robots or robot-like machines. It wasn't until the mid to late 1970s, however, that manufacturers would start addressing this marketplace through the use of kits. One of the first was the Terrapin Co. of Cambridge, Massachusetts. Their Turtle was an outgrowth of the child's computer language LOGO, designed in the MIT laboratories near

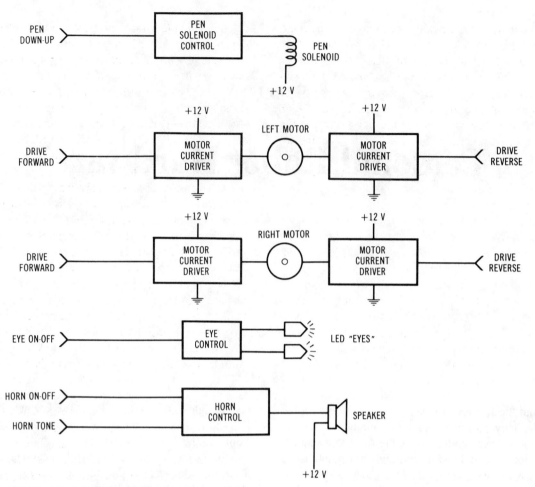

Fig. 5-1. Early Terrapin Turtle functional block diagram.

Cambridge. The Turtle was a two-wheeled vehicle controlled by an external computer through an umbilical cord. There were a couple of LEDs on the front that could blink to give it some personality. There was a small beeper that would also add to its personal capabilities. Also included was a pen solenoid that would allow the rover to draw a trail as it moved about. There was no on-board power system and no on-board brains. A block diagram of the Terrapin Turtle is shown in Fig. 5-1. As you can see, it is not much more complex than the M-2 drive board shown in Chapter 1; however, the ability to purchase a printed-circuit board with components, motors, and everything that fits helped many nonmechanical- or nonelectronic-oriented people to get into the use of robots. Today the Terrapin Turtle is still being sold in a much more refined form for somewhere around $500.

Following the Turtle several smaller attempts were made by various companies. A company called Gallaher Research marketed a number of industrial-like robot-arm kits that did not seem to catch on in the industrial world. One of these was an interesting product called the Grivet Chess Playing Arm that, when programmed and connected to an external personal computer, would allow the arm to play chess with an operator. Around the time this product was introduced, a new horizon appeared on the personal robotics marketplace. The first issue of *Robotics Age* emerged in the summer of 1979.

Robotics Age was an outgrowth of the ideas of a couple of people who worked at the Jet Propulsion Laboratory in California. Allen Thompson, its editorial director, set in motion a means through which hobbyists and industrialists alike could communicate robotic ideas. In an age where there is a magazine for every possible personal electronic entity, *Robotics Age* was a welcome sight. Through the years the magazine has grown, has been taken over by former *Byte* founding editor Carl Helmers, and is now published out of Peterborough, New Hampshire. Most avid robotics experimenters subscribe to the magazine and await its every issue anxiously.

The industry hasn't been without its hard times since its inception, however. In the winter of 1982 two companies announced the availability of a very sophisticated robot rover kit. The first was RB Robot of Golden, Colorado. This product was based on a single-chip microcomputer from National Semiconductor Corp. that allows the user to program the robot in standard Tiny BASIC. It included a wheeled cylindrical body and many add-on options. The second company was the giant kit supplier. Heath Company of Benton Harbor, Michigan. The Heath Co. HERO 1 will be the subject of most of the hardware discussions in this chapter. HERO 1 has outsold the RB robot and all others announced subsequent to that period of time. Around the beginning of 1983 a new robot company, Androbot of Sunnyvale, California, formed by the former Atari executive Nolan Bushnel, announced two robots, one called TOPO, which is a remote-controlled robot that can be hooked to the Apple II computer, and the other called *BOB*, standing for *brains on board*. This was a highly sophisticated 8088-based multiple-processor robot that, as of this writing (which is November 1983) has not become available.

There are several dozen smaller robot kits that have become available within the last six to twelve months. One of the most exciting is an updated turtle-like rover called the Scorpion. This product is manufactured by Rhino Robots Inc. of Illinois. They are well known in the small programmable-arm business and usually provide superior-quality products.

There will be others. Where does the future of the personal robot lead? It should be obvious to anyone reading this book, that within five to ten years personal robots will be as common as personal computers. Several advances will be made in the personal computer industry and power systems and sensor systems that will allow safe and efficient use of robots in the home.

Some of the first applications for the personal robot will be in the cleaning area, which will start as vacuum cleaners, floor cleaners, and eventually get into other tasks. The next section on applications will cover that more fully. As far as the future goes, it should suffice to say that there will be many rovers scurrying around the homes of not only hobbyists but businessmen as well. There will be robots that take out the trash and robots to mow the lawn. Within the next few years large manufacturers will come out with better and better kits and robots.

As I alluded to in the last paragraph there are several applications for home-based robots. The first, of course, is the vacuum cleaner variety. When you study the pattern that a person follows when vacuuming you will find that in many cases it is very regular. To program a robot to go back and forth with a slight left to right or right to left motion in between each forward trail is a relatively simple task. The problems involved in a vacuum cleaning robot are the detection and avoidance of obstacles. Detection of large obstacles such as refrigerators, chest of drawers, beds, or other solid-mass items is fairly easy. Detection of a chair leg, however, because there are four and because they are relatively thin compared to the air mass around them, is much more difficult. Also, the task of programming the robot's reaction to the detection of a pending collision is complex.

From here there is another application that I could foresee a use for. If you are one of the people that throws your dirty laundry into a pile all week long and then must separate it, at the end of the week or twice a week or whatever, by colors or fabrics, you could foresee the need for a robot that would be able to separate the laundry and load the washing machine. I call this a simple laundry robot. It would be placed between the washer and drier, a basket of clothing would be set in front of it, and it would proceed to, through the use of visual and tactile sensing, separate various types and colors of clothing into piles. It could even be outfitted mechanically to load the washing machine with each pile. If you really wanted to get fancy you could build a special washing machine that would only load itself with certain colors and certain fabrics at certain cycle times. I'm sure that in the future the home laundry industry will be evaluating such products.

Another area is in the dishwashing and dish-stacking application. As you know the home dishwasher has saved many a homemaker's hands but still requires the loading and unloading of the unit. There are times that loading the unit can be just as involved as washing by hand. To have a robot that could clear the table and load the dishwasher would be very convenient. Of course, as you can imagine, there are a number of mechical problems that could be approached with a system as complex as that. The placement of the dishes, the food left on the dishes, the placement of utensils on top of dishes, and the general arrangement of the kitchen must be considered before such a system could be implemented.

Of course, there are other applications for personal robots, mostly in the security field. Having a watchdog is a nice thing but watchdogs require food and sometimes they fall asleep. A robot watchdog, on the other hand, sufficiently programmed, could watch a house throughout a whole weekend, with motion and sonic detectors, awaiting any intruder's entry. It would not need food, except for an occasional power recharge and, depending on the design of the internal system, may not require a recharge over a full weekend's work. Another security area would be a watch for fire or flood. Realizing that the smoke detector has cut down the number of fire-related deaths in the world and that robots are much more expensive than smoke detectors, the use of a robot to detect a fire may seem to be overkill. However, utilizing the same watchdog robot as a fire-detecting robot and/or a flood-detecting robot would put all the security issues into one main unit. This robot could be programmed to make decisions based on other inputs during a crisis situation. For example, how many times have you set off the smoke detector when burning a hamburger on the stove? A robot standing nearby, which is realizing that you are cooking this evening's supper, would not signal a smoke or fire alarm in light of the information presented to it.

As you can see, there are many practical uses in the home; these are only a few. For the hobbyist and electronic or mechanical experimenter, there are several uses that come to mind. Path planning and motion research, manipulator research, and human interaction programming research can be carried out with a robot vehicle. You will find that, once you have a maneuvering platform, one that perhaps has the ability to grasp things, finding out how to program a platform like that can take up many months of work. Today, there is still so much experimentation that must be done in order to advance the field of robotics that you still have the possibility of being a superstar in the field.

At this time let's look at the requirements for a home personal robot. By requirements I mean, if you were to have the ideal robot to experiment with, to utilize within your home, what would its characteristics be? You know, in the previous chapters, we had a wish list type of system that would define the basic idea of what we were to design. At this time let's make a wish list for a personal home robot. Let's keep it simple, however and not assume that we could build an android of the future.

The following outlines a wish list that I have put together for a basic research robot.

Wish List
Personal Robot

1. Wheeled base
2. Speech synthesizer
3. Sonic or visual distance measurement system
4. Simple programming language
5. Battery operation
6. Arm with gripper

As you can see, it resembles, very closely, that of the rover requirements in Chapter 1. However, on this wish list you will notice the inclusion of information as to how to program the robot. This item was not included on any of the previous wish lists for the industrial variety. In a home situation you would want to be able to program the robot as easily as possible. A standard computer language such as BASIC might be the easiest implementation of such a programming language. However, you could design an even simpler language that would be even more robot-like for use in programming this beast.

The following takes the wish list and turns it into a specification.

Detailed Specifications
Personal Robot

1. Drive system
   Two wheeled (powered)
   Third castor wheel (or wheels) for balance
   Direct gear drive with stepper-motor control
   Turning radius of 12 inches under special conditions
   Optical wheel position feedback
2. Controller
   Low-power CMOS design
   8-bit microprocessor
   Distributed processing of I/O functions
   Serial interface between functions
3. Programming
   Robot specific language
   English language commands
   Auto teaching pendant mode
   Availability of BASIC-like language
   Program pak carriers
4. Power system
   Individual power centers
   Overall battery monitoring system
5. Manipulator
   Two-fingered gripper
   Elbow with extendable forearm
   2-lb gripper force
   2-lb manipulator life load
   24-inch reach at full extension
   Retractable into body

This specification shows turning radius, size of the body, and if there is a manipulator, the grasp and load carrying capabilities are listed. You will notice that speech synthesis and a rudimentary form of voice recognition are included. I do not consider these outrageous demands. In fact, voice synthesis is almost a necessity in a personal robot system. It gives the operator the ability to program a personality into the robot, which is definitely needed when utilized in a family situation. Also included in the specification list is the speed at which the robot travels. This can be critical, depending on whether it is used in a highly trafficked area of the house or not. One thing you will not find in this specification is that the robot must be able to detect the texture or the color of clothing, clear dishes from the table, or bite the mailman when he comes to the door. These demands would be deemed "application specific."

Let's see if there are any personal robots on the market that match our specifications. The general specifications for the RB5X robot are shown in Table 5-1. Similarly, Table 5-2 shows the general specifications of the HERO 1 robot from Heathkit®. As you can see, both are very similar; however, the HERO 1 robot contains a voice synthesizer and manipulator arm as standard equipment. These advantages, however, are overshadowed by the RB5X's ability to talk in a standard programming language: Tiny BASIC. As I said, for the purposes of discussion, the HERO 1 robot has been chosen to be shown as the design model for personal robotics. Here I will go over the internal design and operation of HERO 1. For those of you who have a HERO 1 robot and would like to gain more information than is contained in the technical manual provided with the robot or you would like to expand the robot, I have written a book called *HERO 1: Advanced Programming and Interfacing*, published by Howard W. Sams & Co., Inc., catalog number 22165. It is available now.

HERO 1 CIRCUIT DESCRIPTION

At this time, let's look back to Table 5-2. Here the general specifications for the HERO 1 robot are outlined. Let's go over them one by one. The robot has six general senses, the first being sound. It can detect ambient sound levels over a frequency range between 200 and 5000 Hz. It can detect up

Table 5-1. GENERAL SPECIFICATIONS RB5X ROBOT

Dimensions:	13″ × 21″
Weight:	10 lbs.
Microprocessor:	INS 8073 (National Semiconductor Corp.)
Programming language:	Tiny BASIC
Storage:	8K with 16K add-on
Interface:	Standard RS232C
Data transfer rate:	1200 baud
Power system:	Rechargeable batteries
Manipulator:	Not standard . . . optional
Speech capability:	Not standard . . . optional

Table. 5-2. GENERAL SPECIFICATIONS HERO ROBOT

Dimensions:	20″ × 18″
Weight:	39 lbs.
Microprocessor:	MC 6808
Programming language:	Assembly language, Robot subroutines
Storage:	4K
Interface:	On-board keypad-display
Power system:	Rechargeable batteries
Manipulator:	Arm with extendable forearm, gripper
Speech capability:	Votrax synthesizer
Sense capability:	Light, sound, range, motion, time

to 256 levels of amplitude of this sound. In the light area its resolutions of ambient light levels is 1 part in 256. This is accomplished by a photo sensor mounted in the head. The sound and light detectors share a common circuit. The ultrasonic ranging abilities are very similar to M2's ability using the Polaroid ultrasonic ranger. Heath, however, has designed their own transmitter-receiver system, which allows them to have resolution from 0.4 inch to 8 feet. There is an on-board motion sensor that is a different-frequency ultrasonic beam that will detect any motion within a distance of 15 feet of the robot. The speech capability uses the Votrax SC01A Phoneme Speech Synthesizer IC. Utilizing this speech synthesis system, any word can be constructed in just about any language using 64 phonemes or sound entities. The last sense that HERO 1 has is the sense of time. A CMOS, battery-backed-up clock-calendar IC is included within the circuitry that allows the robot to keep track of time and date.

From here there are different portions of the robot that are specified. First is the head. The head is mounted on a ball-bearing lazy-Susan arrangement that is turned by a stepper motor that has the ability to rotate 350°. The head contains the sensors for the motion detector, the light detector, the ultrasonic ranging system, and sound detector. There is also a small area on top that allows experimenters to plug in circuits using a solderless breadboard. Located in the center of the head is a hex keypad and six-digit LED display for inputting and outputting programs and data to the robot. A photograph of the head assembly is shown in Fig. 5-2.

Fig. 5-2. Photo of HERO's head.

The arm used in HERO 1 is a mechanized-looking entity that has the ability to rotate about the base of the robot because it is physically connected to the back of the head. Therefore, in the horizontal plane the arm may rotate the same 350°. There is a shoulder motor that allows it to raise up and down 150°. The forearm section is an extender-retracter system instead of an elbow mechanism, which allows it to extend 5 inches. There is a wrist pivot motor that may pivot 90° above and below the axis of the arm. A wrist rotate motor will allow the gripper to rotate 350°, and at the end of the wrist is a two-fingered pincer-type gripper that opens to a 3½ inch width. The load carrying ability of the arm is greater

Fig. 5-3. Photo of HERO's arm.

when the forearm is retracted than when it is extended. During retraction, the capacity that HERO can lift is a 1-pound object. With the arm fully extended the load capacity is 8 ounces. The two-fingered gripper is capable of exerting force of 5 ounces on any object that it handles. You will quickly find that 5 ounces is much too little for any practical use. However, in dealing with paper cups or cardboard items, you can successfully emulate an environment where the robot would be used for useful work. You must realize that the HERO robot was designed as a teaching tool only to demonstrate the capabilities of robot mechanics and electronics. Fig. 5-3 is a photo of the complete HERO arm, wrist, and gripper assembly.

The remaining items on the robot consist of its body and drive mechanism. The body or torso is where all the printed circuit cards for drive and logic reside. The power supply circuitry is also mounted here. At the bottom of the torso is the drive mechanism that consists of three wheels, one of which is powered. The steering wheel, which is the third wheel in the triangular arrangement, is not only powered to pull the HERO forward or push it in the reverse direction but it also includes a stepper motor that allows steering over a 180° span. The two rear wheels are nonpowered and are just pulled along by the robot's steering wheel. The mechanism with which HERO tracks its position should be well known to the readers of this book. A metallic disk with reflective slots marked off in equal increments is mounted on the steering wheel. As the wheel moves forward or backward, an optical emitter/detector bounces light waves off of this disk and counts slots as they go by. This is interfaced into the I/O board and eventually the processor counts slots. We will see this later when we go over the operation of the cards. The drive mechanism itself is very inefficient because of the momentum that the weight of the robot produces when it is moving. There is no electromechanical braking system that allows HERO to stop on a dime.

The power system for HERO consists of four 4-ampere-hour, 6-volt batteries. These batteries are arranged as two completely isolated 12-volt power supplies. There are three batteries mounted in the base between the two idler wheels at the bottom and one mounted opposite the arm within the head for balance. These batteries are rechargeable gel-cell types that through the power supply circuit board may be monitored automatically and recharged through a separate charger box that is included within the robot kit.

At this time, we should go into the operation of the different circuits within HERO. Refer to Fig. 5-4. This is a functional block diagram of the HERO system. It not only shows the internal boards and connections, it also shows some external parts. The first thing that we notice down by the power system outside of the robot dotted line is the charger. As mentioned in the last paragraph, the power system of HERO is made up of rechargeable gel-cell batteries. This charger is a separate power supply-like box that is plugged into an ac outlet, the output of which is plugged into HERO. Recharging HERO's batteries takes approximately 6 to 8 hours depending on how deeply they were discharged. The charger itself provides an approximately 20-volt unregulated dc source to charge the batteries. There is a resistor in series with the output that not only limits the current to the batteries, which prevents damage but also lights an indicator light that shows the amount of charge being taken on by the robot. The human operator can then look at this light to determine when the batteries are fully charged. Because of this resistor, the batteries cannot be overcharged.

From here let's look into the power supply. The power supply is a switching regulator type. A block diagram of its internal operation is shown in Fig. 5-5. More of the charging circuity is located on the board for both battery systems. There are two regulated outputs that supply 5 volts to the logic circuit and 12 volts to the motor circuits. There are also on-board detectors for low voltage that are set to provide a logic level signal to the processor when either of the voltages goes below a set specification. This allows the processor to annuciate the fact that the voltage has dropped and it also feeds back into the regulator circuit, which stops it from drawing current from the batteries. This will protect the batteries from an overdischarge.

From here we move into the processor logic portion of the robot. In the center of the functional block diagram in Fig. 5-4 you will see the CPU and I/O portions. Each of these is a separate printed-circuit board. The CPU contains the Motorola, Inc. 6808 microprocessor and associated bus buffer logic. It also contains 4K of CMOS static memory and an 8K debugger robot language operating system. Also on-board are various input ports and output select lines that connect through the I/O board to the various sense inputs and motor drive outputs. There are ports on-board that interface to the display keyboard assembly and the interrupt processing logic is located on this card.

The I/O card, which is physically plugged onto the back of the processor board to eliminate cabling between the two,

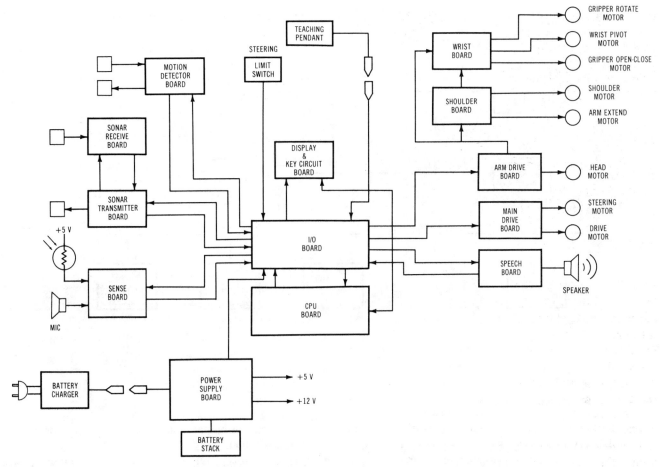

Fig. 5-4. Functional block diagram of HERO 1 robot.

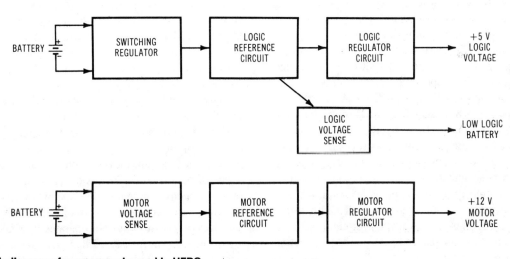

Fig. 5-5. Block diagram of power supply used in HERO.

contains all the drive output buffers that connect to the sense boards. Also on this card is the real-time clock calendar circuit, drive-wheel position sensor logic, and the low-logic and low-drive battery detect switches. Another portion of this board is used as part of the ultrasonic range timing circuit, which consists of a counter and a time-base that determines the distance of the sonar hit. Together these two boards form the heart of the HERO 1 system.

At this time let's take a closer look at the processor logic. Fig. 5-6 is a block diagram of HERO's CPU card. Central to the diagram and to the functionality of the card is the Motorola, Inc., 6808 microprocessor. This is an 8-bit microprocessor that is a member of the 6800 family. The difference between the 6800 and the 6808 is the fact the 08

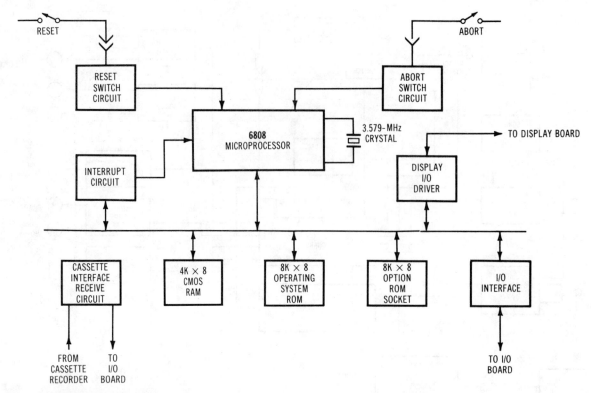

Fig. 5-6. Block diagram of CPU card.

variety has an on-board clock circuit. As you can see, the 3.579-MHz crystal is used as time base for all operations. This gives the processor a 1.1-microsecond instruction-cycle time. The address lines coming out of the microprocessor (there are 16 of them) are buffered and sent on to the other circuits within the system. The data bus, however, is unbuffered. As you can see from the diagram, the system memory with which HERO operates is in two sections. There is 4K of CMOS read/write memory and 8K of what is called an operating system. This operating system will be covered in some detail in the following chapter. The 4K memory is made up of two 2K × 8 CMOS RAMs. This is where the program is stored that HERO will operate on. Also on the diagram you can see the interrupt status register. This is a hardware-produced external interrupt status much like the one discussed for use with the mail-carrying robot. The 6808 allows only one interrupt input. Upon an interrupt, the processor will go to this interrupt status register and read the value present there. This register allows for eight different interrupt requests. I have taken the liberty to show the actual interrupt status register circuit in Fig. 5-7.

As you can see from the figure, the 74LS73s are dual JK flip-flops. Interrupt requests are low-level pulses that, when triggered, will set a selected 74LS73. It is possible, through the use of U411, for the processor to reset any of the requested flip-flops. U412 is used as an 8-bit read-only register. It is through U412 that the status of each of the eight flip-flops will be read back to the processor data bus. As you can see, the combination of U413 and U423B provides a single interrupt output that will call the processor when a request is found.

There are several other circuits on the board referring back to the block diagram of Fig. 5-6. There is the reset and abort switch debounce circuits and the cassette tape interface receive circuit. It is possible to load programs from cassette into the CMOS RAM or dump programs that you have developed out to tape cassette for storage. The circuit used on the CPU card is the standard Kansas City type interface that has been used by hobbyists for years.

What I would like to point out in going over the general circuitry for HERO is the systems level approach taken in the design of the robot. As you can see from the main block diagram of Fig. 5-4, there are separate entities and each performs a function; they are connected together through a series of output ports. This is different than our distributed processing environment that was shown in earlier chapters; however, it is an effective way of building a rover system. The Heathkit HERO system operates in a fairly noise-free environment. The unit that I have has worked perfectly for over a year now without any noise-related problems.

From here let's see the way the I/O board is partitioned within the system. The I/O card block diagram is shown in Fig. 5-8 and mainly consists of many 8-bit output buffers and 8-bit input gates. These are used to communicate with the various sense boards and mechanisms throughout the robot. The buffers are 74LS374 8-bit latches that provide a fair amount of drive capability to power the cables necessary between the senses and the circuitry on the sense boards. In

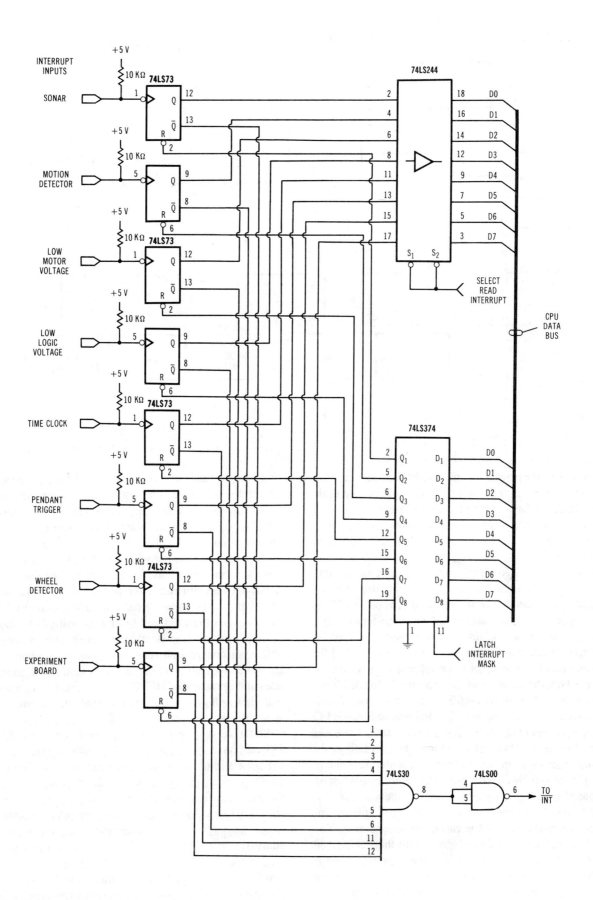

Fig. 5-7. Schematic diagram of CPU interrupt circuit.

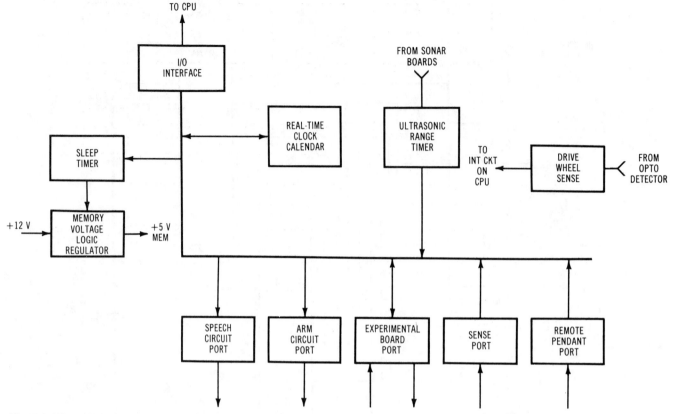

Fig. 5-8. I/O card block diagram.

fact, most of the sense boards utilize CMOS circuitry throughout, which presents a very small input current drain to those latch outputs. Also, as mentioned before, there is the real-time clock. This is an OKI Semiconductor MSM5832 that allows seconds, hours, minutes, day, date, and time to be stored. The operator must set this real-time clock up on first power-on of HERO, and from there on it should stay fairly accurate throughout HERO's lifetime unless the batteries, for some reason, get disconnected. The part itself has its own time base, which uses a 32-kHz watch-type crystal, and the processor, at any time, can read the clock value.

It might be said that this board was kind of a catch-all for smaller items that would not fit on other boards within the system. There are portions of various circuits located here. The speech chip used with HERO requires some 12-volt drive transistors, which so happened to be placed on this I/O card. Also on the card is the ultrasonic range timer that takes the received pulse from the amplifier board located in the head and triggers a flip-flop that stops a counter from counting. We'll go over this circuit in much greater detail when we get into the sense operations. Also on the card are the memory voltage regulators used to supply 5 volts to the memory chips during what is called the sleep period. The HERO robot has the ability of going into the sleep mode where all other circuits are disabled except for the memory. During this time a special 5-volt supply must be available to the memory chips. On this card, 12 volts enters from the motor supply and is regulated by some 78L05 regulators and sent on to the memories through a special sleep timer switch.

At this point you probably have a number of questions. The first of which is, why aren't we going deeper into the HERO circuitry? Another might be, you've told us a lot about its capabilities but what does HERO do? I'm sure that there are many questions like this. All your questions will be answered between the latter half of this chapter and the following final chapter. What I am trying to do by initially showing you the CPU and I/O functionality is to show the systems approach the Heathkit designers took in coming up with a personal robot rover educational kit. After you look at the general architecture of the robot, then I will go into the uses and operation of HERO. Looking at it from this back-end view will give you a better feel for why the architecture was chosen.

So far we've covered the power supply charger, CPU, and I/O card blocks in the functional diagram of Fig. 5-4. Looking back at this figure, let's pick another area. The display and key circuit board shown in the upper portion of the diagram is the only operator interface to HERO. This board, shown in block diagram in Fig. 5-9, provides operators with a hex keypad through which they may enter data and addresses in talking to the 6808 directly. The six seven-segment LED displays will allow address and data to be simultaneously displayed for operator use. Each segment of the six displays is addressable as a separate memory location.

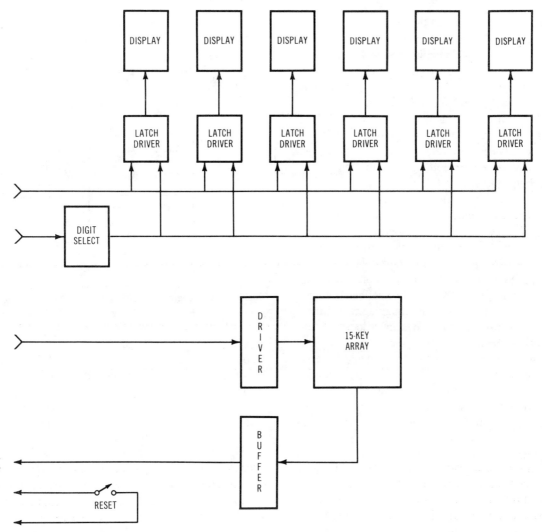

Fig. 5-9. Display and keyboard block diagram.

Therefore, in your applications program, you will be able to utilize the display as part of the I/O mechanism. Each key on the keyboard is also an address location and may be read through an input port on the CPU card. As you can see from the diagram in Fig. 5-4, the keyboard circuit connects directly into both the CPU and I/O cards.

Looking back at the block diagram in Fig. 5-9, we see that the circuitry on the display and key circuit board is very simple. The keys are scanned by the microprocessor in much the same way any keyboard circuit is approached. The displays are addressed and latched as 8-bit entities, seven of those bits being lightable segments and the eighth bit being the decimal point for each display. Later we will go into the exact circuitry of the board. It is then you will see the utter simplicity of this circuit that has been designed as the operator interface.

Let's pause a moment here and look at the systems aspect of the operator interface to HERO. We just covered the basics of the display and key circuit board. You might find it interesting that Heath decided to use a rather dated technique of providing a 16-key hex keypad and six seven-segment LED displays for operator I/O. Although most of today's personal computing machines have done away with the operator keypad, Heath felt that in the educational environment such a simple interface might help to break the uninitiated hobbyist into this very complex field. It also allows a rather inexpensive method for communicating with the processor. Consider the alternatives. A CRT-type circuit like the one used with M2 although not requiring an on-board monitor would require that the person purchasing the HERO either have access to or purchase an additional CRT monitor. This would add to the cost of the robot and also limit its availability to some people. The ability to purchase an all-in-one trainer kit that requires no other purchased items is obviously a plus. In being able to build each entity including the operator interface the student gains an overall appreciation for the project. It also brings the user down to the machine level that HERO operates at. This level is not the only level that the operator is expected to work at. As you will see later in the next chapter, there is a higher-level robot language built into the operating system of the robot.

Before we go any further, let's stop at the functionality that HERO possesses as we have it right now. We have a processor board, an I/O card, some memory, a real-time clock, a power supply, and a keyboard display. Let's find out with just this central processing core, what the architecture that HERO is based on is capable of doing. It is here that some of the software built into HERO has to be discussed. It is this fine line between hardware and software that sometimes is very hard to separate. Let's get into the operation of the HERO debugger.

The word debugger is a holdover from the early microcomputing days where the hex panel and display were operated through a rather crude monitor type program that allowed an operator to enter in programs from the keys, display them, and subsequently change them when a bug was found in the program. The early debuggers only possessed three or four commands. HERO, on the other hand, utilizes an extremely complex and extensive debugging capability. It not only allows you to open and close memory locations, and deposit and change data in those locations through the keypad, but it also allows you to load and store data to cassette tape, and set breakpoints to where the robot will stop operation when a certain address is encountered and numerous other items. Why would you need such a capability in a personal robot? Remember the fact that HERO 1 was designed as a teaching tool. The builder of the HERO kit is being brought along as the next designer of a robot. Getting this builder to the level of thinking that the initial designer of HERO came from will put this new builder in tune with the systems approach to robotics. As I have said many times in this book and in the previous volume, there are many right ways of designing robots. The HERO system is only one of those right ways.

Let's take a quick look at the debugging functions available on HERO. Later, in the next chapter, each command will be gone over much more thoroughly. During this discussion it would be helpful to refer back to a diagram showing the keyboard located in HERO's head for it is through here that all of the debugging actions are accomplished. Fig. 5-10 is a representation of the HERO keyboard. There are several modes of operation that you can put HERO into from the initial turn-on state. After the power is applied, the voice accessory will repeat the word "ready" and the seven-segment LEDs will show the word "HERO" and the version of the current firmware. There are four modes that can be entered immediately from this point. Depressing key 1 at this time will put the robot into what is called the program mode. The program mode is akin to the assembly-language mode of most microcomputers. Depressing key A instead of key 1 will put the robot into the repeat mode, which will allow the robot to actually do or repeat a program. The difference between the two modes is that the repeat mode allows the robot to execute special high-level robot language commands that you will get into in the next chapter. There are

	DO D	EXAM E	FWD F	USER 3
REPEAT	AUTO A	BACK B	CHAN C	USER 2
LEARN	RTI 7	BB 8	BR 9	USER 1
MANUAL	INDEX 4	CC 5	SP 6	
PROGRAM	ACCA 1	ACCB 2	PC 3	
	0	RESET		HERO 1

Fig. 5-10. Representation of keypad on HERO's head.

other keys that will put HERO into other modes on power-up. Key 3 will enter HERO into what is called the utility mode. This mode will provide services that you might want or need during daily operation. The first action in the utility mode would be initialization. Pressing the 3 key for utility mode and then a 1 will immediately send the robot into the initialization routine. This will act as a house cleaning for the microprocessor's various memory locations that contain variables indicating positions of the various motors and other relevant items. Other types of utility commands are the cassette down-load and up-load and the ability to set the time and date of the real-time clock calendar. As I said before, each one of these commands will be gone over more thoroughly in the next chapter.

Another mode of operation is when you initially press key 4. This is called the manual mode. In manual mode you can physically drive the robot around using a hand-held teaching pendant, which is very much like a steering wheel except there are no wheels involved. Fig. 5-11 is a representation of the teaching pendant. As you can see from this figure, it is possible, through switch conditions, to manipulate any portion of the robot manually. Key 7 will perform a similar action on power-up except that all of the manually driven commands will be sorted sequentially as you enter them. This will allow you to program on the fly, so to speak, which is a very powerful command for a personal robot. Someone who does not have the ability to program in assembly language will be able to create high-level robot language commands simply by driving the robot around in what is called the learning mode (key 7).

Assuming that you have a program entered, the debugging commands that allow you to get in and manipulate are shown in a table in Fig. 5-10. Let's go over them very briefly. If you are in the program mode, which was entered by pressing key 1 after reset, you may now do any of the commands shown in the chart. Pressing key 1 again will activate the actions

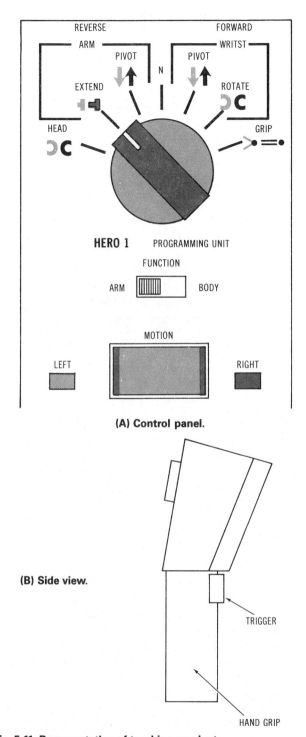

Fig. 5-11. Representation of teaching pendant.

the 6808 is pointing to. Also within the 6808 is a 16-bit index register that is displayed when you press key 4. Whenever the microprocessor needs to perform a logical operation or makes a decision, it consults the condition code register. In some microprocessors this is called a status register. Here the condition code register has bits associated in it that show the current interrupt mask, whether the last operation resulted in a negative condition, a zero condition, an overflow, or what is called a full carry. This is normally indicated as a 6-bit output that is called the condition code register. Pressing key 5 will display the contents of the condition codes in ones and zeros on the display. Looking at the display the leftmost digit indicates the half-carry bit, then the interrupt mask negative, zero, overflow, and full carry.

Microprocessors temporarily store their current address when they are interrupted or when they are told to go off and perform another subroutine somewhere else in memory. They store this temporary information on a thing called a stack. This stack is a push down set of memory locations that are normally pointed to by a stack pointer register. This is also a 16-bit entity located within the microprocessor. Pressing key 6 will display the contents of the stack pointer register. Now from these keypresses it is possible to find out, after you have stopped a program from running, the contents of these different registers. It is beyond the scope of this book to explain the theory of debugging programs. However, for those of you who are familiar with debugging techniques, you will see that there is a nice ability to be able to view the innards of the microprocessor through HERO's buttons.

Going further, there are things called breakpoints that may be set which, when encountered, will temporarily pause the microprocessor in a program. By pause I mean that the 6808 will not totally stop a program and go back into the executive keyboard handler. It will simply put it on hold and go to the keyboard to look for another instruction. It is possible to enter eight breakpoints within your program through HERO's keypad. When a breakpoint is encountered, you may resume the program by pressing key 7. Pressing key 8 is a single step at this point, which will allow you to single step through your program to see each action. At each of the single-step points, the address and data at that address is displayed.

From right here you have the ability to change that data by pressing the C key. Setting a breakpoint is done by pressing key 9. When you press key 9, you will then type in the four-digit hex address that you wish that breakpoint to be at. Whenever you press the reset key all breakpoints are cleared.

Let's assume that you want to enter a program. Pressing key A after key 1 will put you into something called the auto mode. In the auto mode you don't have to keep entering addresses before you enter data. Pressing key A will display a prompt asking you for a four-digit starting address. You enter in the four digits, then you enter in your two-digit hexadecimal value of data for that address. After each two digits, the address display will increment to the next address,

shown on the tops of the key letters. You'll notice above key 1 is the acronym for accumulator A. Within the 6808 microprocessor there are two general-purpose accumulators, A and B. Pressing this key will display the data contents as a two-digit hex value on the seven-segment readouts. Pressing 2 will similarly display the contents of accumulator B. Key 3 will display the contents of the 16-bit program counter, which is the current position that the program counter within

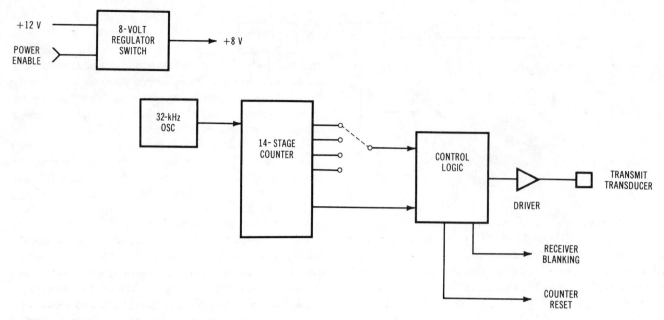

Fig. 5-12. Block diagram of sonar transmitter.

which is called auto load. At any time you can examine memory even from a reset condition by pressing the E key and then loading in the address. You can start examining memory by subsequently pressing F each time you would like to step forward through memory or B each time you'd like to step back through memory. At each of these memory displayed locations you have the chance to change the data by pressing the C (change) key and then entering the next two-digit data. The only other key left to talk about is the D key. When pressing D, you are again prompted for a four-digit hex address. When you enter these four digits, the program will immediately send you to that location and start running. This is called the *do user program button*.

As you can see from the table, we have just run out of possibilities on the debugger. Yes, it is a software issue but I wanted to bring it up to answer the question, what can HERO 1 do? From here we know a little more about this first personal robot. Yes, it is a programmer's tool but if you remember from the learn mode it is also a beginning programmer's tool. So we have satisfied the ability to do things at a higher level without having BASIC or some other known programming language. In fact, HERO can be programmed without knowing any language. So from our wish list, one of the most important requirements has been met. From here let's go back into the circuitry to see how some of the other requirements have been met.

HERO 1 SENSES

What good is a robot if it can't feel its surroundings? HERO has a tremendous number of senses built in. We went over them very briefly at the beginning of this chapter. From here we're going to cover them in a little more detail as far as the functions that each one of its sense boards performs. Once again, we are not going to get into the circuitry details deeply. We are going to look at the architecture of the sense areas. Look back at the functional block diagram in Fig. 5-4. Around the periphery of the CPU I/O core in this block diagram, we see various senses and sense boards outlined. One of the most impressive senses that HERO has is the ability to judge the distance of objects in its path. This is called the sonar transmitter and receive portion or the ultrasonic ranging system. It's broken up into two separate parts. There is good reason for this decision. The receive and transmit portions of a highly sensitive sonar system are very prone to disturbances even from each other! Let's look at the transmitter portion to see how it is accomplished in HERO. Fig. 5-12 is a block diagram of the ultrasonic ranging transmitter. The main purpose of this transmitter is to transmit short ultrasonic pulses through the air that can be received by the receiver portion. These short pulses must be of a precise frequency and duration. Within this board is a 32-kHz oscillator that is user adjustable so that during the installation phase of the sonar transmit/receiver pair, the user can fine tune this sense so that it works properly in the environment. By environment I mean that sonar systems are very sensitive to the type of air, temperature, and ambient operating conditions they are used in. Looking back at the block diagram, we see that the 32-kHz clock feeds a 14-stage counter. The output of this counter is selectable. By that I mean that it is possible to select a number of transmit pulses to be sent on through the control circuitry and out to the output drivers. The output of this 14-stage counter is a series of pulses. You may select through a jumper configuration whether you want

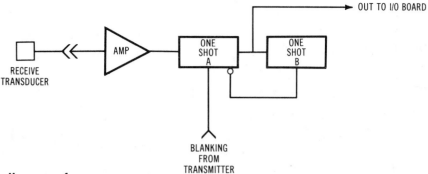

Fig. 5-13. Block diagram of sonar receiver.

as few as two or as many as eight pulses of this 32-kHz signal to be sent as one that is called "ping" of the transmitter output. From here there is control circuitry on the board that will allow that one short burst to go out. During that, the receiver blanking output is active. You would not want the receiver to couple the large energy transmit pulse into its amplifier and, therefore, it is blanked or disabled during the transmit pulse period. The drivers on board, as seen in the block diagram, are to drive the ultrasonic transducer, which is connected in the front of the head. The other circuitry involved on the board is a voltage regulator that takes 12 volts from the battery system and regulates it to an 8-volt level. As you can see, the circuitry on the board is CMOS and does not require that it run at 5 volts. It was found by the designers of HERO that an 8-volt level on this circuitry was much more reliable.

At this point I would like to explain a little bit about the sense boards through the HERO system. Through the I/O board HERO's CPU can address each one of these boards. The way that the designers of the HERO system coped with the enormous power drain of having 14 printed-circuit boards on at the same time was to selectively apply power to whichever sense you are talking to. This can get tricky so within the executive program when you select a sense board, there is an automatic turn on of power through a switch based on each board, and then there is a slight delay in the program to allow that board to stabilize and then the program goes on. To select any of these boards is usually done in the robot mode of program, and this is automatically accomplished. If you do it through assembly language, you've got to put that delay in yourself. So, looking back at this block diagram, you see the switch that allows the 12-volt level of power to enter into the 8-volt regulator on the card. You will see this similar action on all other I/O type cards throughout the system.

Now let's look at the sonar receive section. The receiver is very simple. It consists, mainly, of a very sensitive wideband amplifier. This amplifier, by the way, is the same amplifier that was shown in Chapter 1 for experimentation with wire-guided vehicles. It is a very reliable high-gain-type integrated circuit that is inexpensive. The block diagram of the sonar receive board is shown in Fig. 5-13. Here you can see that the receiver transducer is connected to this wideband amplifier and amplified so that it will only respond to the detected sonic signals within the 32-kHz frequency range. Its output then goes on to a one-shot multivibrator. This one-shot is disabled during the blanking period; as I explained before, during transmit, the output of the one-shot cannot happen. This circuit will only allow one, which would be the first, sonar-received pulse to activate its output. When we get into the actual circuitry of the board, I will give you a better definition of how this is done. As you can see, the output of the board is a TTL-level pulse. This pulse now goes to the I/O card where the ultrasonic ranging timer exists. This timer operates also at a 32-kHz rate. It has been counting 32-kHz cycles ever since the transmitter sent its transmit pulse out into the air. When this first received signal pulse comes in, it will stop the oscillator input to this counter and trigger an interrupt on the CPU card. From here the processor has the possiblity of going out and reading the contents of this counter. Now, obviously, the counter will be some strange value that is not exactly distance. In order to compute distance from this value, you've got to take into account the clock frequency that it was being incremented at and the time it takes a sonar frequency signal to travel through the air.

Going further, the sonar counter on the I/O board is incremented every 31.2 microseconds. It is an 8-bit counter that will give a value of 00 to FF in hexadecimal notation. This yields 255 counts. The number 255 will yield a 7.9 millisecond incremental value. Now it is time to look at how fast sonar waves go through the air. It is known that the frequency of sonar travels 13,080 inches per second or about 1090 feet. Therefore, each increment of 31.2 microseconds will yield approximately 0.4 inch of travel so if you get one count when you read it back, you are a little under a ½ inch away from a detected object. If you read the value FF, you are approximately 8.6 feet from an object. For those of you interested in pursuing the HERO circuitry in greater depth than I have explained and will explain. I mention that there is a book called *HERO 1: Advanced Programming and Interfacing* that can be purchased through Sams, and in this book is a complete table of all of their values and their distance outputs

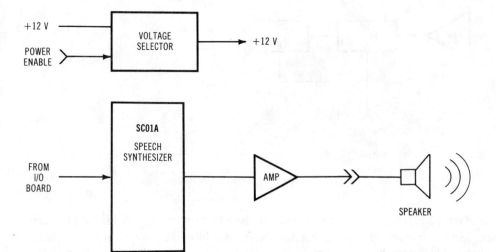

Fig. 5-14. Block diagram of speech accessory.

already computed for you.

Now we've got HERO telling distances, but it's really not "telling" yet. The robot does have the power of human speech or almost human speech. This is accomplished through the voice synthesizer known as the Votrax SC01A. It is a small 22-pin plastic part that has a tremendous ability to simulate the vocal cords of a human. This is done through a series of codes called *phonemes* that simulate different sounds of speech. The speech board is shown in block diagram form in Fig. 5-14. As you can see from the block diagram, there really isn't much to talk about unless you get very deep into this IC. The output of the chip goes directly into an audio amplifier and then off the board to a speaker located in the head. Six-bit digital phoneme code values are fed to this synthesizer chip and out comes speech. It's basically as simple as that. The part runs off 12 volts, and, as you can see, the on/off sense switch switches the 12 volts coming directly from the power supply onto the card. Within HERO's operating system, speech is used extensively to the point where if the logic voltage is low, HERO will tell you in voice: "low voltage"! Every time you press reset, it will say "ready" and you can program it for a number of output phrases. Within the on-board ROM are several canned phrases that can be used, most of which you would want to use in a demonstration situation. Having the ability for your personal robot to speak, as I said earlier, personalizes it much more. When you program the robot you should try to use the speech as often as possible. This will greatly enhance the user interface.

From here, let's go on to another sense that in itself is very reliable and actually somewhat remarkable. The ability to sense light in various quantities and to sense sound can be used in many applications. Both of these senses are performed with one circuit card. Unfortunately, both cannot be used at the same time because of the multiplexing technique. Fig. 5-15 is a block diagram of what is known as the sense circuit card. This board has two inputs from the outside world. One is a simple, light-dependent resistor or photocell.

The other is a small speaker just like the one used for the speech output except it is being used here as a dynamic microphone. Both of these connect into the sense board through discrete amplifiers made up of many transistors. From here they become a digital logic signal that varies from 0 to 5 volts. This varying voltage is then fed to an 8-bit analog-to-digital converter. This will net 256 defined levels of input value; therefore, depending on which is selected, either the sound or the light detection, the output of this converter will show 256 amplitude levels of input. Also on this board is an LED driver that drives eight displays. The displays are small lamps located on the board. These are used to determine values before they get up to the microprocessor. It's another one of those extras that Heathkit usually puts into its kits. You can set up this board without having a microprocessor within the system. It will show you the output of the A to D so that you don't have to read it on the 8-bit output. This may be selectively enabled or disabled by a manually switched jumper. Once again on the block diagram, you can see that the 5 volts going into this board is switched by the enable signal when the board is addressed.

From here we go to another interesting sense. This one utilizes ultrasonic waves in a different way than the ranger. The motion detector, shown in block diagram form in Fig. 5-16, uses ultrasonic pulses in a closed-loop system. What it does is flood the area within 15 feet of the robot with a 35-kHz ultrasonic constant output signal. As long as that does not change, in other words, nothing disturbs the air around that 15 feet, the output is inactive. As soon as the time frame for those pulses coming back into its on-board receiver changes, it will immediately interrupt the processor that some motion around it has occurred. This works exactly the same as ultrasonic burglar alarms that are available for houses. Looking at the block diagram we see that there is a 35-kHz oscillator whose output immediately goes to the transmitting transducer. The receive transducer comes in and is amplified. There is a low-frequency modulation detector and then it goes on to the bandpass amplifier. This is made up

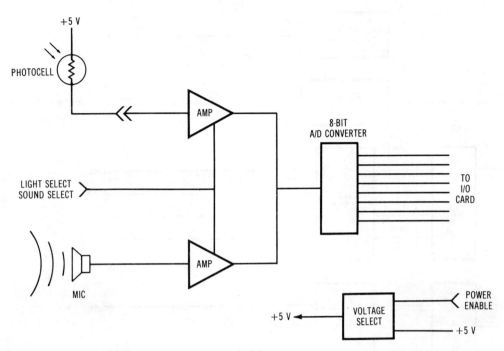

Fig. 5-15. Block diagram of sense board.

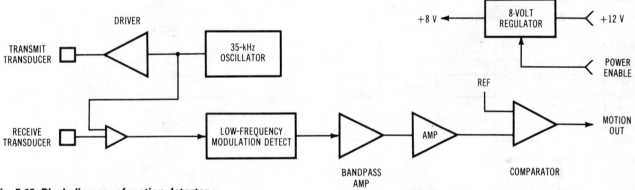

Fig. 5-16. Block diagram of motion detector.

of discrete parts not the same IC that is used on the sonar receive board. And then there is a comparator to see if anything has changed. Once again, the board runs on 8 volts, and it is switched on and off by the processor.

As far as senses go the only sense left to talk about is the sense of touch. In actuality HERO cannot feel through its arm or legs. In fact, HERO's legs are wheels. However, there are boards associated with the arm and drive motor operations. The arm-drive circuit board is shown in block diagram form in Fig. 5-17. Most of this board consists of optically isolated inputs and driver transistor outputs, one for each phase of winding for all the motors, but you'll notice that there are twice as many motors in the arm as are shown on this driver output board. You'll also notice that the outputs are multiplexed. How can they be multiplexed to a motor? Well, it turns out that you can only do half the motors at any one time. The way they worked it was that there are two switches on board that allow the common windings of either the wrist rotate, shoulder and head motors, or the wrist pivot, and gripper and extend motors to be activated at any one time. The processor selects either-or and then the digital drive signals coming from the I/O board to each one of these motors is optically isolated to eliminate noise, as we went over earlier, before this signal is sent on through a driver transistor on this board. The motors are operated at 12 volts. This 12 volts is physically isolated from the 5-volt supply, and this is part of the isolation, as you see here on this card. The drive motor circuitry is much different. Fig. 5-18 is a block diagram of the main drive circuit board. This board has a number of optically isolated inputs that drive four motor windings for the steering motor. From here its similarity with the arm-drive circuit board stops. As you can see in the block diagram, there is a pulse-width modulator that controls the speed of the drive motor. HERO's drive motor is enabled by a series of short pulses. It is known that this is a very effective speed control method—by not fully turning on the motor or

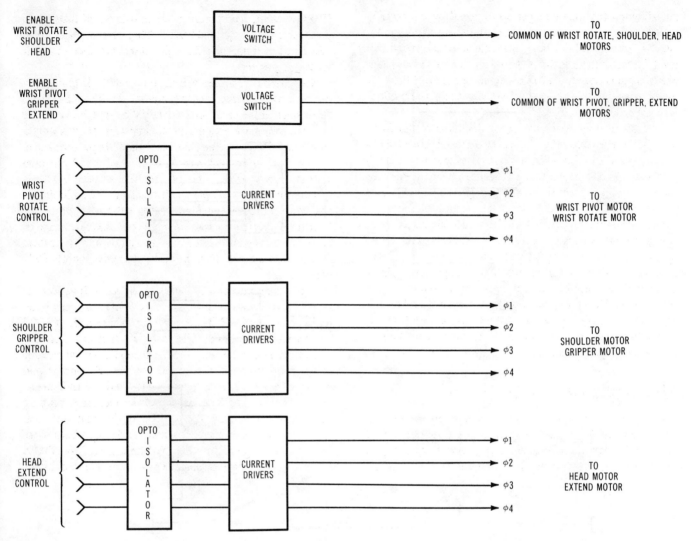

Fig. 5-17. Block diagram of arm-drive board.

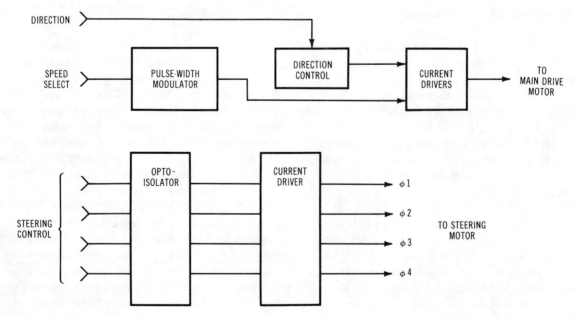

Fig. 5-18. Block diagram of main drive board.

not allowing the motor to get up to speed by temporarily taking away its excitation voltage. The pulse-width modulator is fed with a 7-bit input that selects which speed you want the motor to turn. Six of these bits determine the speed; one of them is an output enable and the eighth bit of the 8-bit motor drive word that comes onto the card selects direction, forward or reverse.

That about does it for the block diagrams of the cards involved in HERO. Let's reflect now on the abilities HERO has as a personal robot against the abilities that we wished a personal robot to have. It has mobility. It has a variable speed; therefore, you may select whichever speed at which you would like to see a robot scurrying around your home. It has a multifunction arm which can be used to pick up objects; although light as it may seem, it still can be used for lightweight objects. The motion detector serves one of the security functions, and the ultrasonic ranger will help it navigate. The light and sound detector can be used in various applications. Depending on how you program the robot, the light detector may be used to differentiate colors from an object that is held up to its sensor. You may find an application in the laundry area.

ADVANCED HARDWARE

Right now, let's digress somewhat from the HERO description and explore the general architecture HERO used but in a totally different context. If you were to design a personal robot utilizing the same functional blocks as the HERO system, incorporating the knowledge of distributed processing that we gained back in Chapter 1, it might be an interesting exercise to go through. At this time, let's embark on such a project.

Fig. 5-19 is a redrawn functional block diagram of a personal robot resembling the internal structure of HERO. The names of the cards are basically the same; however, as you can see, the functional processing core outputs only a single serial data stream and receives a single serial data stream. Each entity within the system is powered by its own separate power system, and the only link between I/O cards and the processor is an optically isolated bus. This eliminates any noise problems and functionally isolates the processor from the rest of the system. It also allows for easy upgrade of other functions on this serial bus. Taking it one step further, Fig. 5-20 is a block diagram of this personal robot's CPU core.

In the diagram you see basically the same functions as HERO possesses. The 4K of CMOS memory has been replaced by an 8K CMOS memory chip. There is 8K of ROM and various bus buffers. The only I/O link is a standard UART. The output of the UART goes to an optical isolator driver. This eliminates the need for the I/O card. Upon elimination of the I/O card, you might wonder what happens to the real-time clock calendar and the ultrasonic ranging timer and other functions that were handled on that card. You will see as we go along throughout the design where these functions get picked up. The real-time clock calendar is located on this processor board. I have gone one step further

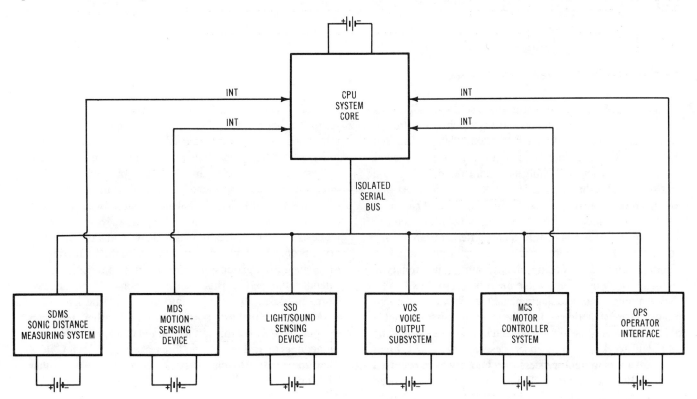

Fig. 5-19. Personal robot functional block diagram.

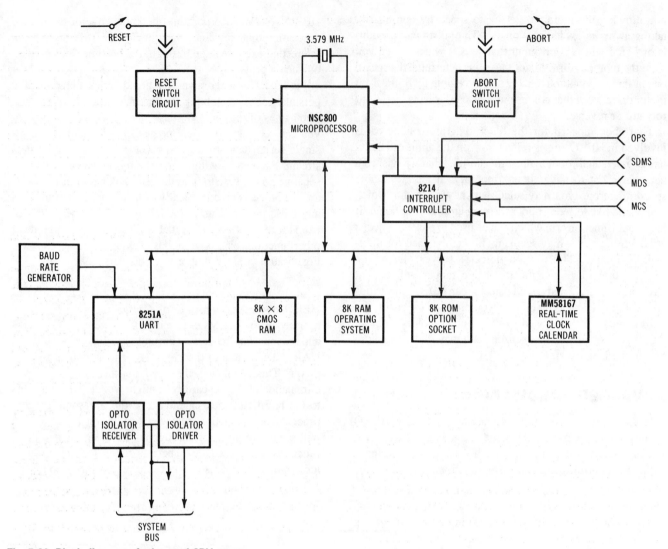

Fig. 5-20. Block diagram of advanced CPU core.

and shown in Fig. 5-21 a semi-detailed schematic of such a board. As you can see, the circuitry is much more streamlined as advances in LSI have been used. Also the fact that there is no I/O card to interface with makes this a much more compact arrangement. Commands from the processor are sent out serially through the UART to all of the SCU distributed processors on the bus. There is only one I/O location outside of the processor board. Within the board, as you can see, there is a clock calendar IC that is different from the one utilized in HERO. The one chosen here is manufactured by National Semiconductor Corp. It is a further generation type chip and easily interfaces to the microprocessor bus.

Examining the block diagram and the schematic, the first thing you must notice is that the microprocessor is not a 6808. This was chosen by desire and not as a necessity. The 6808 will work very well under these circumstances; however, advances in microprocessor technology have occurred since HERO was designed. The IC chosen for the advanced CPU core here is the National Semiconductor Corp. NSC800. This microprocessor is built in CMOS, which drastically cuts down the power necessary for operation. Also, the fact that the NSC800 emulates the powerful Z-80 microprocessor architecture adds a certain amount of flexibility in the software design. Coming off of the microprocessor, you'll notice an LSI interrupt controller. This interrupt controller, denoted on the diagram as the Intel 8214, does all the functions that the circuit shown in Fig. 5-7, but does it all with one IC. This allows eight interrupt requests and automatically masks them out as per their priority. From here, looking around the diagram, we also recognize the microprocessor-compatible real-time clock calendar, the 58167. This is also a National Semiconductor Corp. IC. We will be going into that a little bit later. The RAM program memory portion, as I said earlier, has been extended to 8K. It turns out that in today's age of advanced semiconductors, it is possible to purchase an 8K × 8 CMOS static memory. This part is packaged in a 28-pin dip that takes up less room than the two 24-pin parts used to provide 4K on HERO.

There is an 8K × 8 ROM slot for the executive program. If

Table 5-3. Interrupt Code Chart

Priority Request		RST	D_7	D_6	D_5	D_4	D_3	D_2	D_1	D_0
			1	1	$\overline{A_2}$	$\overline{A_1}$	$\overline{A_0}$	1	1	1
Lowest	$\overline{R_0}$	7	1	1	1	1	1	1	1	1
	$\overline{R_1}$	6	1	1	1	1	0	1	1	1
	$\overline{R_2}$	5	1	1	1	0	1	1	1	1
	$\overline{R_3}$	4	1	1	1	0	0	1	1	1
	$\overline{R_4}$	3	1	1	0	1	1	1	1	1
	$\overline{R_5}$	2	1	1	0	1	0	1	1	1
	$\overline{R_6}$	1	1	1	0	0	1	1	1	1
Highest	$\overline{R_7}$	0	1	1	0	0	0	1	1	1

you are using the NSC800, obviously, you will not be able to utilize HERO's executive ROM. However, if you decide to build a circuit like this and provide it with a 6808 microprocessor, then it is possible to utilize the ROM here. Last but not least is the UART that connects it with the rest of the robot circuitry, and that is an old standby part, the Intel 8251A. The operation of this part is dealt with in detail in *Microprocessor Based Robotics*.

From here it would be helpful to go into the operation of the NSC800 to give you a feel for what it takes to interface, hardware-wise, to the rest of the circuitry. Looking at the schematic diagram in Fig. 5-21, we see that the clock generator for the NSC800 is built right into the chip. If we were to use an NMOS Z-80, this would not be the case; however, all it takes is the connection of a crystal to the X-in and X-out pins of the NSC800 to provide it with oscillation capabilities. The address bus of the NSC800 is multiplexed. The lower 8 bits of an address are shared with the 8-bit data bus. An address latch enable signal on its falling edge will latch the low-order addresses; therefore, as you can see on the schematic, an 8-bit latch is used to provide these address outputs. There is a separate read and write line coming from the NSC800 that fits into the system very nicely. This separation of read and write functions is necessary for use with the 8251A and the real-time clock calendar. There are several levels of interrupts that the NSC800 can handle without the use of an external interrupt controller. This allows for extra interrupt inputs besides the eight that are already taken up. Looking at the diagram, the interrupt controller (8214) is connected to the standard interrupt input to the microprocessor. The 8214 allows eight different interrupt requests. When one of those eight comes in, an interrupt request is generated to the microprocessor. At that point the microprocessor will read the output of the 8214, which will give it a special code that tells which one of the inputs has interrupted. The codes per interrupt are shown in Table 5-3. In fact, at this time, a memory map of the processor core board should be examined. A complete map of the locations of all of the devices on the board is shown in Fig. 5-22. You will notice from this map that not everything on the board is a memory location. The UART that talks to the SCU-controlled I/O boards is an I/O location. The NSC800 allows for a separation of I/O addresses and memory addresses, which virtually extends the capabilities of the microprocessor. This allows 65,536 memory type locations, and I have chosen to utilize memory locations for both the real-time clock and the interrupt controller. It also allows 256 I/O locations of which, at this point, we only need 2. The remaining portions of the NSC800 resemble standard microprocessors. The reset input is rather standard. The reset output is an output that goes to the rest of the board to reset the various devices. At this time let's look at the real-time clock calendar and its operation.

Realizing that the clock calendar is a set of memory locations, Table 5-4 lists the various memory locations and their functions within this integrated circuit. It turns out that there are 23 locations used within memory. Each one of these locations, as you can see by the table, has a function. Looking at the table now we see that the first eight locations are the actual counters used for seconds, minutes, hours,

Table 5-4. Memory Locations — Functions of Real-Time Clock

ADDRESS	FUNCTION
C000	Counter — thousandths of seconds
C001	Counter — hundredths and tenths of seconds
C002	Counter — seconds
C003	Counter — minutes
C004	Counter — hours
C005	Counter — day of week
C006	Counter — day of month
C007	Counter — months
C008	Latches — thousandths of seconds
C009	Latches — hundredths and tenths of seconds
C00A	Latches — seconds
C00B	Latches — minutes
C00C	Latches — hours
C00D	Latches — day of week
C00E	Latches — day of month
C00F	Latches — months
C010	Interrupt status register
C011	Interrupt control register
C012	Counter reset
C013	Latch reset
C014	Status bit
C015	"GO" command
C016	Standby interrupt

days of the week, dates of the month, and months. From this you can get an idea of the power of this chip. The only item not included in the clock calendar is the year. The next eight locations are latches that allow you to write data in for seconds, minutes, hours, days of the week, dates of the month, and months. The interrupt output of the 58167 allows eight possible interrupt actions. When you latch a certain time, date, etc., it is possible to compare this to one that comes up during normal operation. In this way you can basically set an alarm clock mode. When the two match, an interrupt can be generated. You also have the ability to produce interrupts every tenth of a second, every second, every minute, every hour, every day, every week, or even once a month. Those are the eight possible interrupt applications that this part can be programmed for. In timing applications, you can see the power of being able to get an interrupt every second. The complete chip is operated using a 32-kHz clock crystal very similar to the one used with the HERO processor. The device is fabricated in CMOS technology and operates off the battery supply that is located on the board

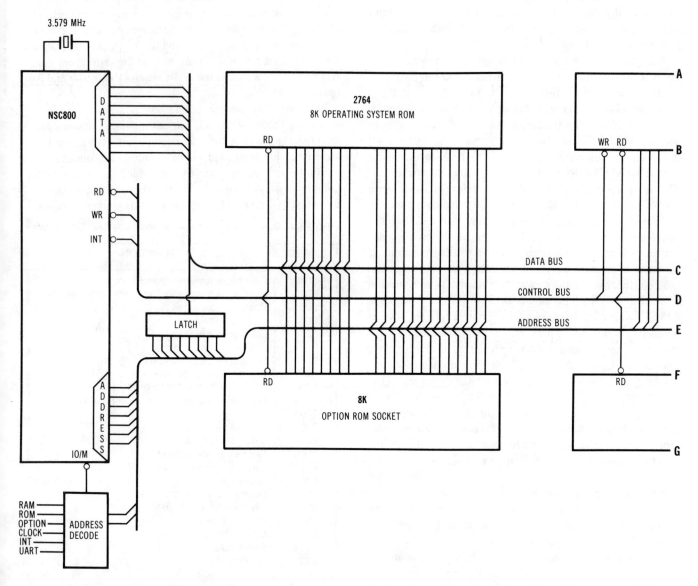

Fig. 5-21. Schematic of robot CPU core card.

with it. In this application I have decided to provide a separate power supply battery for the clock calendar. In this way, if the main robot batteries are disabled or discharged, it will not affect the clock. If you calculate out the size of the battery and maximum drain from the clock, you probably will never have to replace this battery for the lifetime of the robot.

Looking back at the table, there is an interrupt status register and an interrupt control register. The interrupt control register is a mask register that regulates which of the 8 bits in the status register go as an interrupt. The status register can only be read. It contains the present state of the comparator, which compares the counters and latches and the outputs of the tens of seconds, seconds, minutes, hours, etc., counters. These are 1-bit outputs that are on when the counter increments. This part has the ability to enter a stand-by mode whereby the rest of the circuitry is shut down, the part is in a power-down condition, and the only interrupt enabled is called the stand-by interrupt. This standy-by interrupt is enabled by writing a 1 on the low order bit line with the stand-

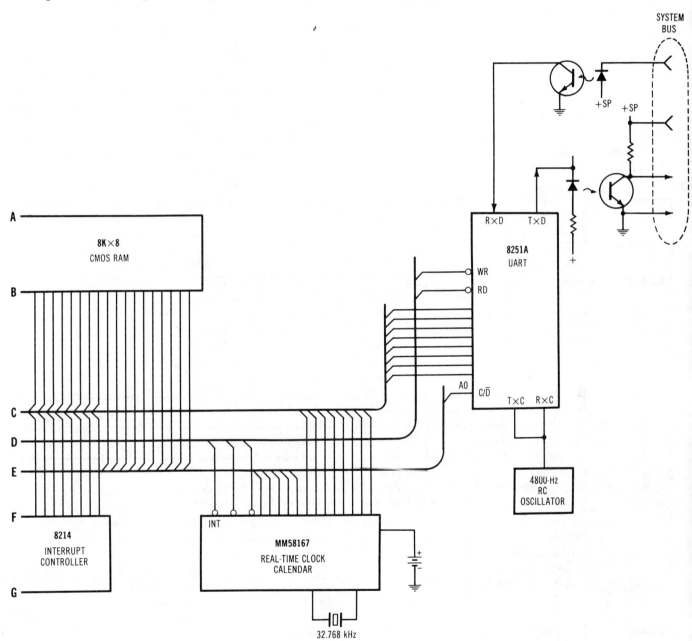

Fig. 5-21. Schematic of robot CPU core card.

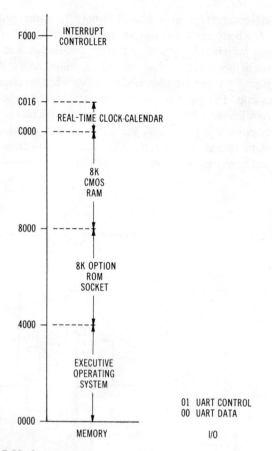

Fig. 5-22. Core processor memory map.

between the two is the fact that all operations must be commanded through the UART, and other circuitry, not shown on the card, is the cassette interface or the ultrasonic ranger/timer. You will find in other areas of the advanced core architecture these functions are taken care of. Speaking of the architecture, let's move along through the UART to the various I/O portions of this advanced core robot.

Looking back at the overall block diagram of Fig. 5-19, let's go to the far left and pick up the new sonic distance measuring system that carries the acronym SDMS. Fig. 5-23 is a block diagram of this newly enhanced measuring system board. Comparing this with the block diagram of the sonar transmitter and sonar receivers of Figs. 5-1 and 5-13, you will notice that most of the circuitry in those other diagrams is included here. Also on here is the intelligent SCU controller I/O commander portion. The ultrasonic range timer is built in to the SCU; therefore, no external circuitry is necessary.

Going further into this board, Fig. 5-24 is an actual schematic of the SDMS. As I explain the sonic transmitter/receiver circuitry, let it be known that these are adaptations of the actual circuits used in the HERO robot; therefore, they are tried and true and can provide you with an insight into the operation of HERO as well as the steps necessary to enhance such a system. As you can see in the schematic a 32-kHz oscillator is devised by using a single gate in an RC oscillator-type configuration. The 10-kΩ pot that provides feedback from the output back to the input of the gate allows you to set the frequency of operation. This 32-kHz clock period goes into the CD4040 CMOS counter. This counter in conjunction with the dual flip-flop 4013 provides 14 stages of counting. The control circuitry shown using the CMOS quad two-input NOR gate 4001 will perform various timing functions within the system. Bear with me as I explain the operation of this circuit. The 32-kHz oscillator not only feeds the 4040 counter but also feeds one of the gates in this array. The output of this gate goes directly to the drivers for the transmit transducer. The other input to this gate is controlled eventually by the output of the counter. The counter will provide eight pulses of 32-kHz oscillation and then shut off this gate. Actually those eight pulses are selectable. During the time the transmitter is transmitting, the receiver blanking output is enabled, which will disallow any crossover feedback from the transmit to the receive transducer. Looking now at the receiver portion of the schematic, you can see the RCA CA3035 wide-band sonic receive amplifier, which is set to receive the 32-kHz signal. The output from the amplifier will trigger one stage of the dual one-shot. When this stage is triggered, the output to the interrupt logic on the processor board is sent out as a constant width pulse. The second stage of the one-shot, however, will then trigger and disable the ability of the first stage of the one-shot from triggering on a second echoed pulse. This will stay in that condition until a reset signal comes in from the transmitter. This reset signal is the opposite phase of the blanking signal;

by interrupt address selected. On the next counter/latch comparison the output will turn on. This could be used to implement your own sleep mode within the robot. You could have the rest of the robot power down awaiting a certain time or date and the clock calendar stand-by interrupt output could then enable the power to start up the robot when that time exits.

Setting the clock is very simple. You just write to the counters. Setting an alarm condition is done by writing to the latches and enabling the alarm condition interrupt. Using the GO command, as shown in the table, will start up the clock. Whenever you want to read the current time, you read the counter values. When you do read the counters, you should read the status bit location. This will tell you if a counter was changing state when you read it. It will give you the ability to reread in order to eliminate any possible errors. That about does it for the real-time clock. As you can see, the circuitry required to interface it to the NSC800 is very simple.

The memory components on the board and other memory address locations are selected through the memory address decoder ICs shown on the schematic. The memory ICs, as I said before, are single-chip implementations of static ROM and RAM parts. They read and write like standard memories and require no further discussion. What you see in the schematic of Fig. 5-21 has approximately the same power as the HERO robot CPU and I/O cards. The major difference

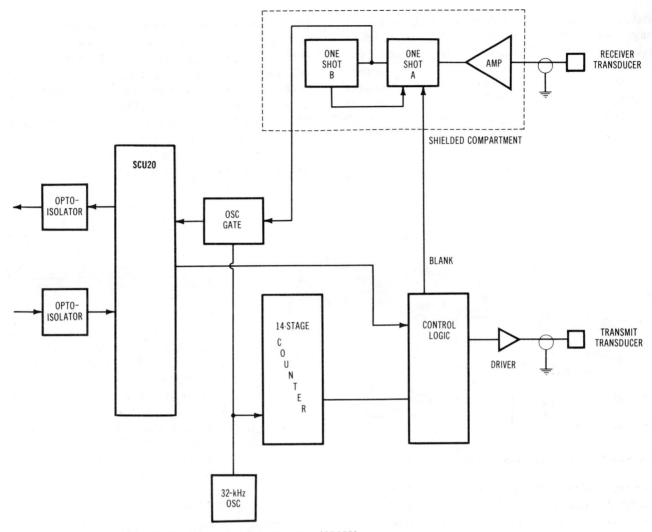

Fig. 5-23. Block diagram of sonic distance measurement system (SDMS).

therefore, this circuit will only allow the first actual echo to interrupt the processor.

Where does the SCU come in? If you will notice the 32-kHz clock also enters the counter input to the SCU. Logic shown on the card will allow this oscillator to enter the internal SCU counter as long as the transmitter has transmitted a pulse and the received echo has not been received. As soon as the received echo enters the board, the input to the SCU counter is shut off. After the interrupt is received at the processor board, it is now possible for the processor to query the SCU20 and read the counter value. In this way two signals have replaced the more than eight lines of interfacing between the processor and I/O sections of the HERO robot to read the 8-bit counter value. If you were to determine that the ranging system should be implemented in a multiple-transducer arrangement, it would be just as easy to read multiple counters with the same SCU. In the original HERO system that would be a major modification to the circuitry.

Moving along in our discussions, the motion-sensing device, MSD, is shown in schematic form in Fig. 5-25. This system does not require an SCU. It is the same circuitry that is used on the HERO robot. Looking at the schematic you can see the 35-kHz oscillator in the upper left-hand portion of the figure. The receive transducer will be constantly picking up this 35-kHz signal that will be reflected off of objects within its 15-foot range. If one of the reflecting objects moves, it will result in a variation in the amount of energy that's reflected back to this transceiver. This variation will be picked up as a low-frequency amplitude modulation. The low-pass filter, made up of a combination of LM324 op amp sections, will detect this modulation and cause a high-level output that determines that motion is appearing in the vicinity. The schematic is relatively straightforward. The difference between this system and the HERO system is that there is no power on/off switch controlled by the processor. Its output goes directly to the CPU core interrupt inputs.

The next place an SCU is used is in the light/sound sensing device or just known as the SSD. Shown in block diagram form in Fig. 5-26, the schematic of which is shown in Fig. 5-27. Here the implementation is somewhat different than that in the HERO project. Both the sound and light amplitudes may be read without regard to selecting either-or, because a multichannel analog-to-digital converter is provided for this purpose. The A/D converter itself is interfaced to an SCU and read through the input port. The amplification circuitry is also much more simplified. It is still only possible to read a variation of 256 levels of amplification on each input. As you can see from these figures, the ability with one address to talk to either the sonac distance measuring system or the light/sound sensing device, cuts out a lot of circuitry and makes the software much more streamlined.

The voice output subsystem shown in block diagram form in Fig. 5-28 also utilizes an SCU device. I did not select the Votrax voice synthesizer for this design as a newer 5-volt

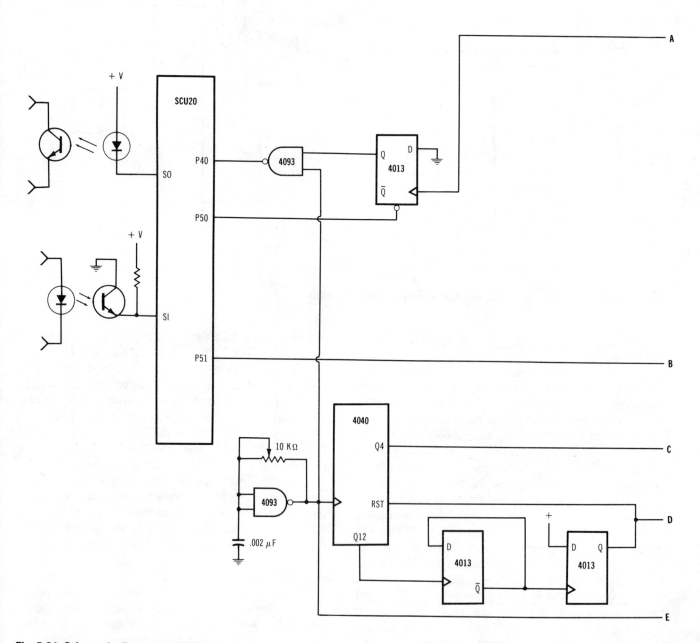

Fig. 5-24. Schematic diagram of SDMS.

device has arrived at the marketplace. This part, manufactured by General Instruments, is a synthesizer that utilizes the allophone principle of speech emulation. The allophone is a further extension of the phoneme and allows for much more natural sounding speech. There are also 64 allophones that can be used to produce any type of speech output you desire. At this time look at the schematic in Fig. 5-29. As you can see, the SCU part talks to the voice synthesizer IC, which is denoted as the SP-0256-AL2, through the use of six allophone code lines. Also, there is a single busy line and a single load line. The proper method for loading in allophone codes is to check the status of this busy line to determine if the processor is outputting speech or is in the input mode. If the speech processor is not busy, then a toggle from high to

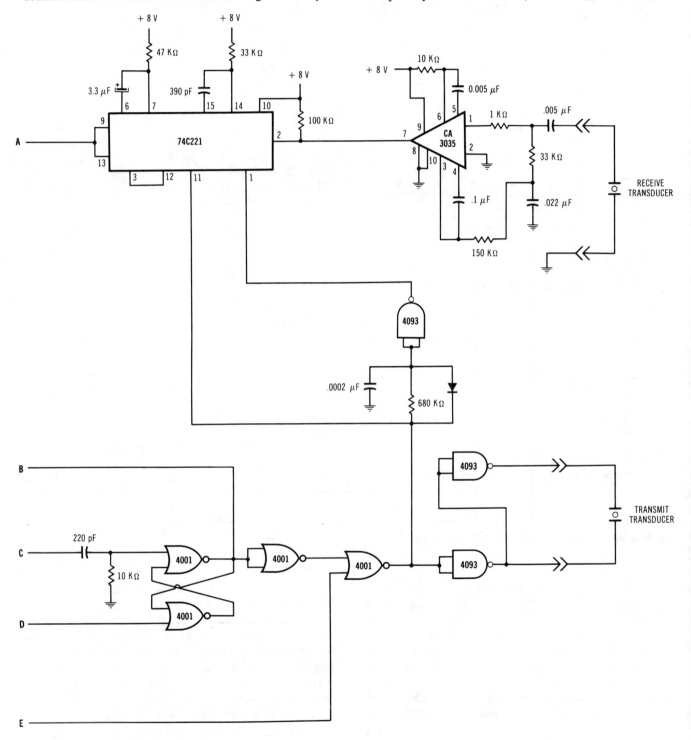

Fig. 5-24. Schematic diagram of SDMS.

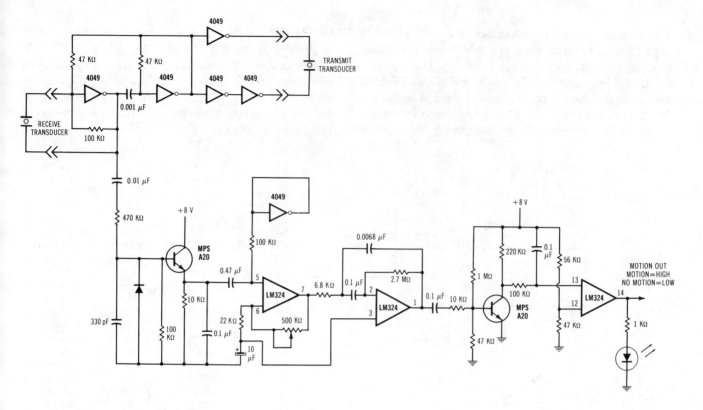

Fig. 5-25. Schematic diagram of motion-sensing device (MSD).

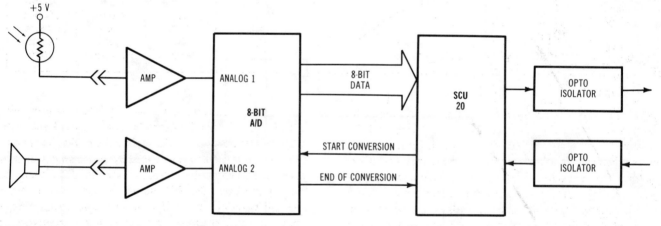

Fig. 5-26. Block diagram of light/sound-sensing device (SSD).

low and then back again of the load line is required with the 6-bit allophone code on the data lines. Table 5-5 shows all of the allophone codes used by the General Instruments part. To talk to the world the CPU core must send allophone codes to the SCU. It must also read the status of the busy line and toggle the allophone code load line. As you can see on the schematic in Fig. 5-29, the output of the allophone speech synthesizer is amplified by the audio amplifier LM386 and sent out to an external speaker. One of the bigger differences between the Votrax part and the General Instruments part is the fact that the General Instrument part does require a crystal that adds some cost to the system. Shown here is the 3.12-MHz crystal necessary for operation.

Moving along quickly, the next board in the system is the motor control subsystem or MCS. This board also utilizes an SCU as the interface device. As you can see from the block diagram in Fig. 5-30, a tremendous amount of commonality is employed between this one board and the drive motor controller and the arm motor controller boards of the HERO system. However, when it comes to schematic commonality, there is no connection except in the drive-motor portion. I have utilized the Sprague UCN4202A stepper-motor con-

Table 5-5. Allophone Code Chart

A1-A6 Values	Allophone	Sample Word	Duration	A1-A6 Values	Allophone	Sample Word	Duration
0	PA1	PAUSE	10 ms	20	/AW/	Out	250 ms
1	PA2	PAUSE	30 ms	21	/DD2/	Do	80 ms
2	PA3	PAUSE	50 ms	22	/GG3/	Wig	120 ms
3	PA4	PAUSE	100 ms	23	/VV/	Vest	130 ms
4	PA5	PAUSE	200 ms	24	/GG1/	Guest	80 ms
5	/OY/	Boy	290 ms	25	/SH/	Ship	120 ms
6	/AY/	Sky	170 ms	26	/ZH/	Azure	130 ms
7	/EH/	End	50 ms	27	/RR2/	Brain	80 ms
8	/KK3/	Comb	80 ms	28	/FF/	Food	110 ms
9	/PP/	Pow	150 ms	29	/KK2/	Sky	140 ms
A	/JH/	Dodge	100 ms	2A	/KK1/	Can't	120 ms
B	/NN1/	Thin	170 ms	2B	/ZZ/	Zoo	150 ms
C	/IH/	Sit	50 ms	2C	/NG/	Anchor	200 ms
D	/TT2/	To	100 ms	2D	/LL/	Lake	80 ms
E	/RR1/	Rural	130 ms	2E	/WW/	Wool	140 ms
F	/AX/	Succeed	50 ms	2F	/XR/	Repair	250 ms
10	/MM/	Milk	180 ms	30	/WH/	Whig	150 ms
11	/TT1/	Part	80 ms	31	/YY1/	Yes	90 ms
12	/DH1/	They	140 ms	32	/CH/	Church	150 ms
13	/IY/	See	170 ms	33	/ER1/	Fir	110 ms
14	/EY/	Beige	200 ms	34	/ER2/	Fir	210 ms
15	/DD1/	Could	50 ms	35	/OW/	Beau	170 ms
16	/UW1/	Top	60 ms	36	/DH2/	They	180 ms
17	/AO/	Aught	70 ms	37	/SS/	Vest	60 ms
18	/AA/	Hot	60 ms	38	/NN2/	No	140 ms
19	/YY2/	Yes	130 ms	39	/HH2/	Hoe	130 ms
1A	/AE/	Hat	80 ms	3A	/OR/	Store	240 ms
1B	/HH1/	He	90 ms	3B	/AR/	Alarm	200 ms
1C	/BB1/	Business	40 ms	3C	/YR/	Clear	250 ms
1D	/TH/	Thin	130 ms	3D	/GG2/	Got	80 ms
1E	/UH/	Book	70 ms	3E	/EL/	Saddle	140 ms
1F	/UW2/	Food	170 ms	3F	/BB2/	Business	60 ms

trollers that drastically reduce the amount of processor intervention when stepping the various motors throughout the system. If you want a better description of how to utilize these motors, look back to the last two chapters. The drive-motor controller portion, however, utilizes the same stepper-motor drive system. I have opted for a stepper motor to give a greater accuracy to its movements. Speeds are accomplished by varying the pattern outputs and step speed to the UCN4202A. You may select one of the three speeds in this system through the SCU. The SCU is also utilized to read the position of the drive motor. This is done by connecting the wheel emitter/detector that we talked about earlier through to one of the SCU's counters. This method of counting position was explained fully in the use of position feedback systems with the SCU in Chapter 3.

The last board in the system is the operator interface. This particular board utilizes the keypad display arrangement very similar to the HERO product except that the display provides for an alphanumeric capability of eight digits. Looking at the block diagram in Fig. 5-31, you can see that the hex keypad of 16 digits is performed by a keyboard scanning circuit. At this time look ahead to Fig. 5-32 at the schematic of the board. All the scanning that is done by HERO sequentially, line by line, through the microprocessor, is done automatically by a single-chip CMOS keyboard encoder. These outputs go to the SCU and can be read by the microprocessor as key entities without the necessity for constant scanning. In fact, in this case, the SCU shown on the board is not a standard Mostek Corp. SCU20. It is a preprogrammed 8748 single-chip microprocessor that is specially programmed for use with operator I/O. Keypresses entered in by the operator are automatically transferred down to the core processor without losing one entry. The alphanumeric display is handled by the core processor sending an 8-byte ASCII message to the operator interface subsystem. This ASCII message is then displayed on the screen. Having the capability of showing alphanumeric digits enhances the operator interface, which I have said many times is the most important aspect in a robot system.

The alphanumeric displays are also advances in electronics. They are made up of two four-digit intelligent display controller LED ICs that are made by Litronics and National Semiconductor Corp. These parts look, electrically, like random access memories to a processor, each

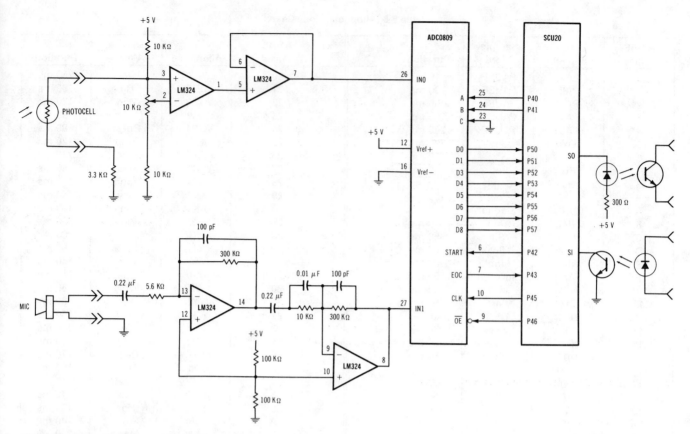

Fig. 5-27. Schematic diagram of light/sound-sensing device (SSD).

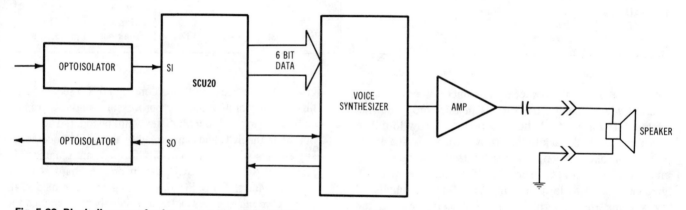

Fig. 5-28. Block diagram of voice-output subsystem (VOS).

have four separate addresses. Internally, they also have a character generator and memory locations. A block diagram of one of these parts is shown in Fig. 5-33. You can see from the diagram that all the necessities of displaying coded messages are included on chip. These devices, fabricated in CMOS, work off of a single 5-volt supply and draw very little current.

Throughout the design of this advanced core processor personal robot, low-power CMOS has been utilized. You have seen the application of the SCU in various modes and the application of a single-chip microprocessor to emulate an SCU. The purpose of this exercise was to show a different approach to the same product, HERO 1. Are there better approaches? You bet there are! There should be ways that the operator interface and the voice synthesizer automatically interact. The key here is to keep the processing down and build automatic controllers that handle it naturally. The motor control subsystem can be enhanced and the sensing systems should be augmented by other senses of sight and tactile feedback. In the next chapter we will investigate advanced software methods for controlling personal robots. We will look at the operating systems of the HERO robot, design a specific operating system for the advanced core processor personal and look ahead at a robot system that alters its operation depending on choices you make when you first power it up.

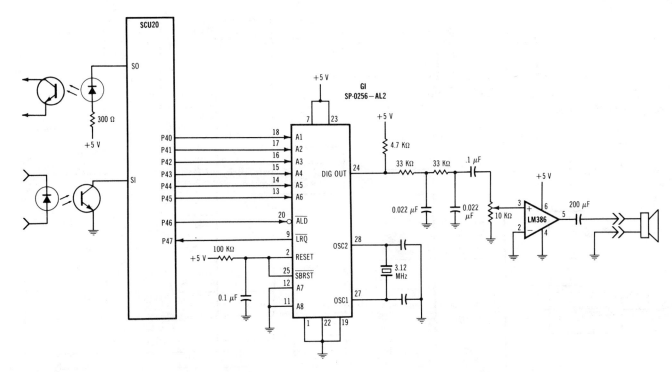

Fig. 5-29. Schematic diagram of voice-output subsystem.

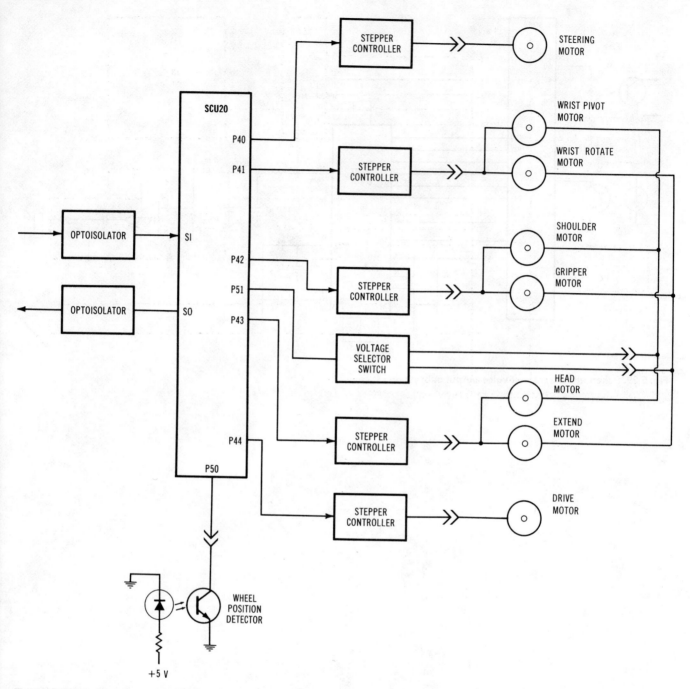

Fig. 5-30. Block diagram of motor-controller subsystem (MCS).

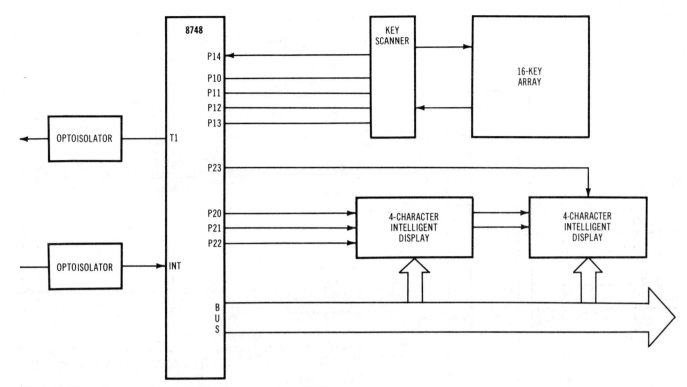

Fig. 5-31. Block diagram of operator-interface subsystem (OIS).

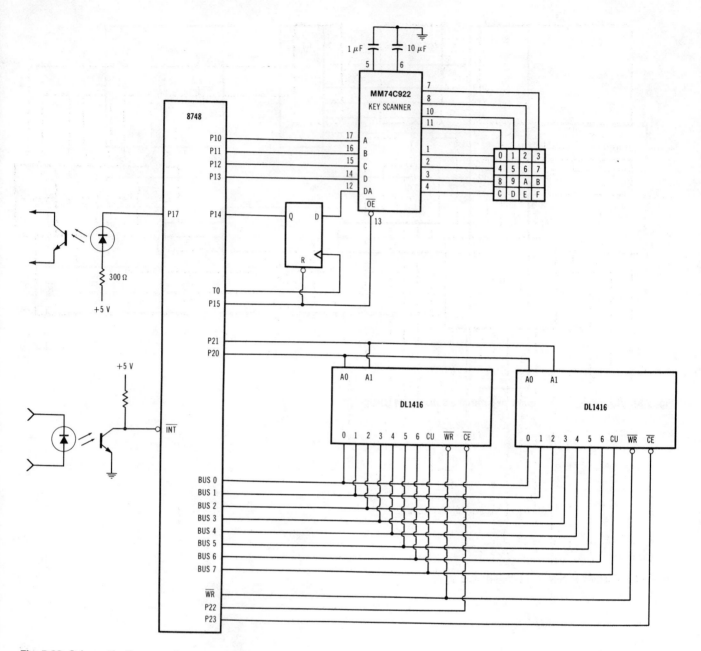

Fig. 5-32. Schematic diagram of operator-interface subsystem.

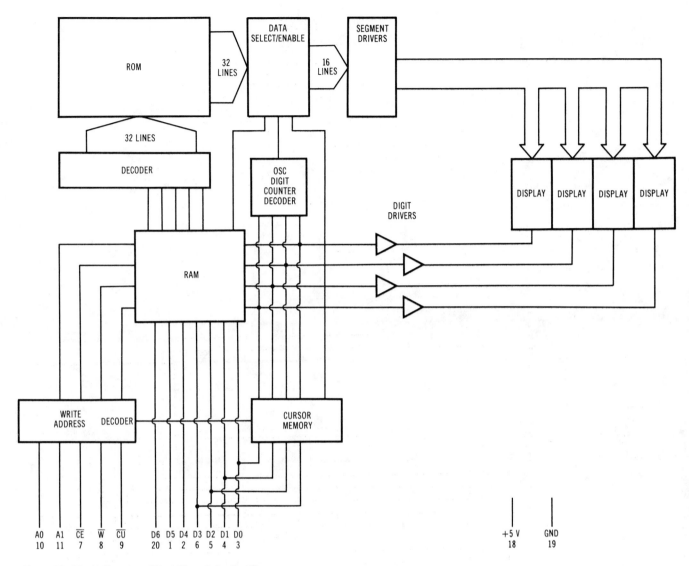

Fig. 5-33. Block diagram of intelligent display IC.

Chapter 6

Personal Robots: Software, Applications, Advanced Techniques

As the name of this chapter implies, we will be filling in the gaps, which pertain to software, left in the last chapter on personal robotics. This chapter is the last one of this book and will serve to pull together all of the subjects that we have discussed within the last five chapters. We will begin with a discussion of robot software in general as it pertains to personal computers, the different language considerations, distribution of language and the media involved, we'll move on and explain the HERO executive system in somewhat more detail without getting into the actual listings of the executive ROM, from there an in-depth discussion of the design of the core processor advanced software that is implemented on the personal robot CPU and I/O structure that was designed in the last chapter and, finally, a section on artificial intelligence as it applies to robotics in general. Specifically in the personal robot area, I will go over a few applications that pertain to self-learning hardware and software. We will look at a path calculation technique for a personal robot vacuum cleaner and revisit the manipulator simulator design and specify a higher-level structure for accomplishing a block's moving program. Throughout this chapter I will try to stress the systems aspect of the complete field of robotics as this is the main thrust of this book.

ROBOT SOFTWARE

Software in general is the manipulation of various hardware items, in concert, to accomplish a specific task. Software is designed in various levels allowing single-task structures through complex, multitask environments. Robot software, in particular, is a new field. Although the CPU, I/O, and memory structures of personal and industrial computing techniques are roughly the same, the overall robot system dictates a specific type of software. Programmers of today are learning new ways of structuring a systems approach to this problem. When you look at the various problems or areas of difficulty that we have uncovered through the last five chapters, you begin to appreciate the role of the programmer in defining the actions of a mechanical beast that must find its way out of a corner guarded on all sides by inanimate objects. It is this programming that gives an animation to the machine that we sometimes call intelligence.

Personal robots, as I stated in the last chapter, must also possess a unique quality of being the replacement of or in addition to man's best friend. This quality can only be achieved through clever programming. Robot software for personal robots must be specific in the human interface tasks that it is to accomplish. Let's look at various languages that can be used to not only create software for personal robots but also languages used for neophyte programmers who work with robots.

BASIC

Standard BASIC, that is built into most low-end personal computers and is available for all personal computers, is by far the most well-known language. With its various English-like statements and easy line-numbering system, even the least experienced of programmers can find themselves designing somewhat advanced programs in no time at all. Several robot manufacturers have included BASIC as part of their product line. The RB5X robot, developed by RB Robot

Fig. 6-1. Photograph of RB5X robot.

in Golden, California, is a robot that is specifically designed around a single-chip microprocessor that executes BASIC as its inherent command set. Fig. 6-1 is a photograph of the RB5X. Looking at this robot, you can see that it is wheeled much like the HERO variety, and there is a series of what appears to be bumpers around the perimeters of its cylindrical body. These are contact sensors that tell the microprocessor when the robot has encountered a physical obstacle. Also built into this system is an ultrasonic measurement device based on the Polaroid ultrasonic ranging kit. It is not the purpose of this paragraph to go into the hardware of the RB5X, but looking at this robot and the fact that its creators decided to standardize on BASIC as its command language should be noted. There are many subsets of the standard Dartmouth BASIC language that are small enough to fit into the internals of a microprocessor. This particular robot utilizes Tiny BASIC, which has an integer arithmetic package that cuts down on the size of the memory needed to implement the language. It also has a smaller set of commands than a standard BASIC. This does not imply, however, that it is any less powerful when it comes to operating I/O, which would be important in a robot subsystem. It turns out that this particular Tiny BASIC is very I/O intensive.

Should BASIC be the personal robot language of the future? Probably not; however, at the present time there are no extensive robot languages or languages designed for robotic tasks available for personal robots. A quick look-through of the various statements found in most BASICs will give you an idea of the adaptability of BASIC to robotic tasks. Following is a list of the standard BASIC statement keywords.

BASIC STATEMENTS

PRINT
INPUT
FOR-TO-NEXT
IF-THEN
LET
GOTO

Although there are only six entries in the table, you will find that probably every implementation of BASIC allows these constructs. Looking at the first line in the list, the PRINT is usually the first statement taught in BASIC classes. It tells the computer to type out a word or a number on the screen or other output device that it uses. In robotics, PRINT might be supplanted by an OUTPUT command. Usually, there are no screens or printer outputs available on robots. If they are, then they are superfluous to the general needs of the robot control system.

Going further down the list, the INPUT normally looks for keyboard entries from an operator that might be entered after a prompt given through a PRINT command. Keyboard entries are another construct that are typically not used in a command robot setup. INPUT may, however, be modified to allow for operator control using other devices. Therefore, a modified INPUT statement may turn into a LISTEN statement if a voice recognition system were to be connected or INPUT may be used as a tactile sensing method such as INPUT from touch sensor or INPUT from motion sensor. However, with these constructs, the word INPUT would become cumbersome; therefore, other methods must be used or other keywords developed.

The FOR-TO-NEXT construct is an internal BASIC use statement, along with IF-THEN-LET-GOTO, that probably could be left intact. You will find that the logic that these statements perform will be universal between personal computers and personal robots. To expand on the table, one might add the Boolean operations of AND-OR-NOT to the constructs that already exist. One could go even further by adding some memory I/O such as disk or cassette access to it. This brings us to an interesting question in personal robotics, that of distribution of software and the media it is stored on. We will go over that as soon as we finish the language discussions.

As you can see from this brief review of the BASIC language portions, the statements used so commonly in all BASIC dialects are indeed adaptable to robotics; however, a great deal more need work. Later, we will look at the development of a robot language that may answer the desires to enhance an operator interface better than the standard BASIC languages allow.

Other standard programming languages, such as Pascal or FORTRAN, suffer the same consequences. Both, with their WRITE statements, run into difficulty where there are no screens or teletypewriter outputs to print to. These languages, although more structured and more powerful than BASIC, can be used, given the same enhancements that we touched on for BASIC. Special versions of Pascal and FORTRAN could be developed to utilize the skills of a robot, however, other languages are more suited toward the automatic control needs of robotics. The languages of LOGO and FORTH tend to suit the needs more reasonably than these other programming languages.

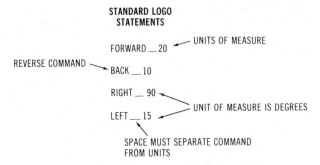

Fig. 6-2. Standard LOGO statements.

Other Languages

LOGO, although designed for children, is a powerful robotics-type language. It is used by elementary and junior high schools to teach the principles of geometry to children. This is done by moving an object called a turtle on the screen of a computer output. This turtle, like the Terrapin Turtle we visited earlier in Chapter 5, is controlled by giving simple English language statements of forward, reverse, right, and left commands. There is a very simple set of preprogrammed commands in the LOGO language, if you deal only with the physical properties of the turtle. Fig. 6-2 shows the command set and the arguments required after the keywords. Looking at this figure, we see that the forward command requires a number of units that represent distance to move in that direction. The turtle, given a command to forward 50, would proceed to move 50 pixels in a direction on the TV screen. In the real-life world of robotics, this may indicate 50 passes of a reflective object past an emitter/detector on the wheel. It may represent 50 times some number of reflective objects. As you can see, the ability to direct an entity, in this case the turtle, in simple English language motion statements, makes the language LOGO very adaptable to robotics.

FORTH is another language much like LOGO. There are some very simple commands built into the language that allow you to enter your own constructs. FORTH is based on the ability to use machine-language commands embedded with high-level calls. These calls are entered from a keyboard, originally as English words, and then you further define them as a subset of machine-language commands. Remembering back to the core processor system that we discussed in the last chapter, it is necessary to develop a great number of machine-language commands with which to move the motors and communicate with the SCU I/O structure. Simple English language statements, such as OUTPUT, may be constructed into a very complex machine-language routine using the language FORTH. Serious personal robotic designers should investigate both languages for incorporation into future projects.

At this time, utilizing the knowledge that we have of the various programming languages we've all used over the years and what we just investigated as far as the possibilities of utilizing these languages with standard robotic use, let's design a full-function language, by defining the constructs and statements used. This language we may simply call ROBOL.

Robot languages may be defined as systems, just as easily as pieces of hardware may. As is the case with any system, a list of requirements should be compiled before the design has begun. Here is such a list for the ROBOL language:

REQUIREMENTS FOR ROBOL LANGUAGE

INPUT FROM: Vision system
              Microphone
              Ultrasonic ranger
              Tactile sensors
              Light detectors

OUTPUT TO: Motors
              Speech synthesizer
              Manipulator

SPECIALS: Loop instructions
              Program storage instructions

As you can see from this list, special requirements must be met in order to implement an effective robot control language. Looking at the list, we see that input commands from various sensors must be specific in their callouts. This is to say, if you want the robot to be able to listen through a voice recognition system, you might use the word LISTEN. Vision systems may use the keyword SEE. A voice output peripheral might do well to be commanded by the keyword SAY or SPEAK. A similar type keyword might be used for an ultrasonic ranging system. The word SCAN, long range or short range, may prove useful. Depending on the exact design of your robot system, the number of keyword statements may grow or reduce from a standard programming language. The requirements set forth in this language list show the basic functions necessary to implement a rover system of the complexity of HERO 1. Output commands such as MOVE, LOWER, LIFT, ROLL, OPEN, or CLOSE might be utilized for motor movements and other manipulator-type commands. Storage commands such as LOAD and SAVE may be allowed to be comprised of the same keywords used in a standard language. Going further in our description, following is a list of keywords that I have compiled for the ROBOL language:

STANDARD ROBOL COMMAND STATEMENTS

SCAN
MOVE
LEFT
FORWARD
REVERSE
RIGHT
PULSE
REPEAT

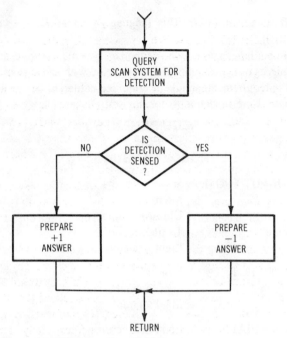

Fig. 6-3. SCAN logic.

HOLD
LOWER
LIFT
PITCH
YAW
ROLL
OPEN
CLOSE
SEE
COMPARE
ROTATE
SAY
LISTEN

Each keyword defines a specific function within a robot system. They should be high level enough so that they will work for any particular hardware implementation. It would be up to the designer to build in the appropriate I/O drivers necessary to support that high-level command.

In assuming a new standard language for robot rovers or personal robots, the design of ROBOL may prove to enhance the programming capabilities of a multitude of people, but does this robot language serve the needs of the uninitiated nonelectronics-type person? Would this type of person be involved in personal robotics? It is a possibility, as time goes on, that neophytes to the engineering world might become involved in utilizing robotic devices. Certainly, if they are to flourish in the home, this would be necessary. Therefore, an even higher-level language must be devised, or some method for compiling ROBOL as a result of a higher language must be defined and examined. Before we get into this higher level, let's finish up on the ROBOL concept. We will call this the standard that all robot commands will link to. It would be necessary that versions of ROBOL be developed for many microprocessor families. It is the purpose of this chapter, and this book, to show a systems approach to the design of software and hardware. No actual microprocessor listing will be shown for the ROBOL language. However, detailed implementations and flowcharts of each structured keyword will be presented.

Starting off with Fig. 6-3, we are going to investigate each keyword involved in the ROBOL language. The first word is SCAN. SCAN, shown in Fig. 6-3, would be a statement used to produce an output that tells a computer logic structure whether an object is within the path of the robot or not. This is akin to an ultrasonic ranging system; however, it is not restricted to any one type of scanning method. SCAN would be called a function in another language. The use of SCAN, as shown in the figure, is called, as an entity, to be placed into a standard variable. A = SCAN would produce either a positive number such as 1 or a negative number such as -1. In the case of a negative number, A would represent the fact that an obstacle has been detected. Utilizing this command, within a program would be very much like the following use of the SCAN statement:

```
FOR HEADPOS = 1 TO 360

OBJECT = SCAN

IF OBJECT = -1 THEN WALL = HEADPOS:
GOTO_____

NEXT HEADPOS
```

Here, the head of a robot is rotated 1° per pass of the program through a 360° rotation. Each time the head is incremented 1°, the scanner is looked at and its output is placed into the variable called OBJECT. In the next line the logic IF OBJECT = -1, meaning if there is indeed a detection of an object, THEN WALL = HEAD POSITION; therefore, at that particular head position, a wall or object has been detected. It would then get out of the program. If not, then it would go to the next head position. High-level logic structures, such as this, are very simple, when you have a routine as simple to understand as SCAN. Looking back at Fig. 6-3, you see that the possibilities of using SCAN with a subset, such as SCAN 0, SCAN 1, SCAN 2, etc., will allow for many different types of scanners on the robot to be called at that time.

In the same manner as SCAN was approached, the statement SEE will allow a robot to enter in data into the programming logic flow. SEE could be designed as simply as looking for an object such as the SCAN routine or could be as complex as actually entering a pattern of data that would be a pixel representation of the area in front of the robot. That was my idea for the SEE routine, as is shown in Fig. 6-4. SEE will allow the robot to enter a 32 × 32 byte pattern into memory. This pattern can then, later on, be compared to detect objects that it has been trained to look for. SEE, of course, would be a

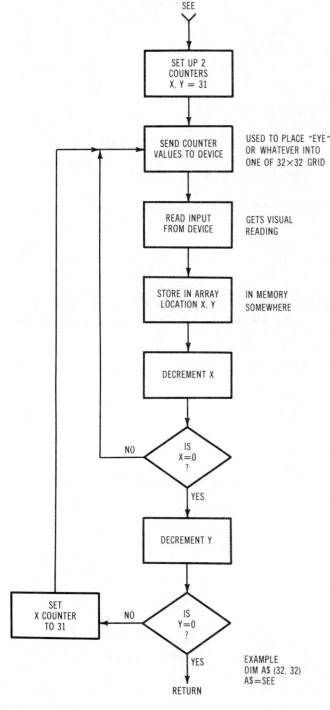

Fig. 6-4. SEE logic.

SAY, as shown in Fig. 6-5, utilizes the same argument as standard PRINT routines. When programming, the programmer would type the word SAY, then in quotations the ASCII string or the variable denoting a string or number would then be spoken. This would require a somewhat high-level text-to-speech algorithm developed for the particular microprocessor involved. These are not uncommon at this time, and it may be necessary to utilize an external piece of hardware to accomplish this; however, the use of the ability to simply type out the word to be spoken will greatly enhance the use of SAY and programmers who are not willing to program at the phoneme or allophone level may find it so easy that they will utilize it more often than not. This would greatly enhance the operator I/O relationship to the robot system.

Complementing the SAY routine would be the LISTEN statement. LISTEN would be used with a voice recognition system, no matter how complex it may be. LISTEN would act much the same as an INPUT statement in standard BASIC. Realizing that there are several different types of voice recognition equipment on the market and in the future many more types will be available, it is assumed, with the LISTEN command, that the output of the recognition system will be a string of ASCII characters. These characters may be, somehow, placed in variable storage or a block of memory that the LISTEN would then draw from. The logic utilized in the operation of the LISTEN command is shown in Fig. 6-6. As I said, realizing that the future holds many new paths for voice recognition systems, it is important that the LISTEN command be designed to accommodate all types. A simple dialog between user and computer or user and robot showing the use of LISTEN and SAY is given here:

```
100 SAY "PLEASE TELL ME THE CORRECT
TIME"
110 A$ = LISTEN
120 IF A$ = "NO MATCH" THEN GOTO 110
130 TIME = VAL (A$)
140 SAY = "THANK YOU"
```

Here you see the statements SAY and LISTEN used in an actual application. As you can see from the dialog, the SAY command could be used as the prompting method for getting LISTEN inputs. It may be possible, within the constructs of the system, to allow for LISTEN to be used with a keyboard or voice recognition subsystem. In that way, the user may opt for a more private conversation using the keyboard instead of speaking into a microphone. This would also depend on the construction of the actual robot.

Moving from the operator interface to the real world of manipulation and control, there is a series of statements used for this purpose. At this time, let's concentrate on the manipulator portion of the robot. You might find the easiest place to start would be at the business end of the manipulator or the actual hand or gripper. It is obvious, to even simple robot structures, that the gripper must have the ability to open and

very complex arrangement of hardware and software techniques that might or might not be possible to do reliably in today's technology. The use of SEE and two other statements will be shown in a later figure. We will come back to the SEE command at that time.

In the last chapter, the ability for personal robots to speak gives them a certain human quality that should always be strived for. The SAY statement allows the programmer to develop a PRINT-like command that the robot would speak.

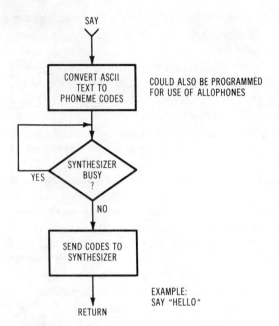

Fig. 6-5. SAY logic.

Fig. 6-7 shows the logic and syntax used when specifying the OPEN and CLOSE statements. As you can see from the syntax, the variables requested after the statement would really depend on the particular system you are implementing. In the HERO robot system there are known hex digit amounts that you may command the gripper to open and close to. These are shown on the figure for reference. To get a better idea of how this ties into the HERO-type system, Fig. 6-8 shows a representation of the HERO gripper and how the action of an OPEN and CLOSE command, which states a known distance value for measurement, is used.

Of course, manipulation commands can be expanded from the OPEN and CLOSE of the gripper to things like LIFT and LOWER for the shoulder motor. The logic used in both of these statements is shown in Fig. 6-9. As you can see, there is very little difference between OPEN and CLOSE and LIFT and LOWER. In fact, Fig. 6-10 shows the syntax and logic used for the PITCH, YAW, and ROLL commands that command the wrist action of a manipulator. These also work on the variable argument system of distance specified after the keyword. Similar, as shown in Fig. 6-11, is the logic behind the EXTEND and RETRACT commands. If your particular robot system does not have an extendable arm, then a subset of the LOWER and LIFT commands could be

close. OPEN and CLOSE may be too general a term to use; however, units may be added to these basic statements, which allow you to open or close a certain known amount.

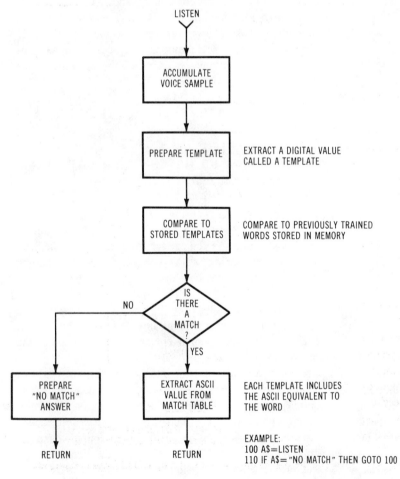

Fig. 6-6. LISTEN logic.

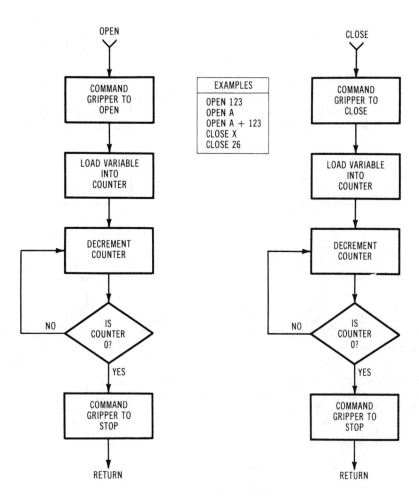

Fig. 6-7. OPEN and CLOSE logic.

used; possibly a LOWER 0 or 1 or A LIFT 0 or 1, depending on whether it's the shoulder or the elbow stated.

From this brief look at manipulator-type commands, you can see the power of having an English language construct used in specifying motion or the direction of motion. When a programmer is programming in this type of command, it is very easy to see a mistake that might be made or to be able to understand the operation of the robot more fully.

From here, let's move back into the vision capabilities of our future-type robot. Go back to the SEE command that we initially talked about in Fig. 6-4. There are two other commands that complement the SEE command in its operation. The COMPARE and ROTATE statements help the robot quantify what it has found in the SEE command. If you will remember, SEE, in my implementation, will send a 32 × 32 byte representation of an image that may be scanned from the front of the robot, or wherever the front of the robot's head is pointing, in order to determine if this array or pixel information, called the image, exactly or somewhat exactly represents a known image that might have been stored earlier. There must be some way of comparing the two arrays. The COMPARE statement, shown in Fig. 6-12, will perform this match up.

You will notice from the COMPARE syntax that a vari-

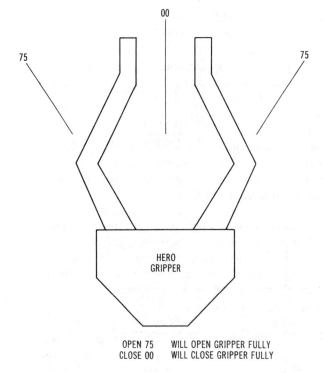

Fig. 6-8. HERO gripper used with open and close statements.

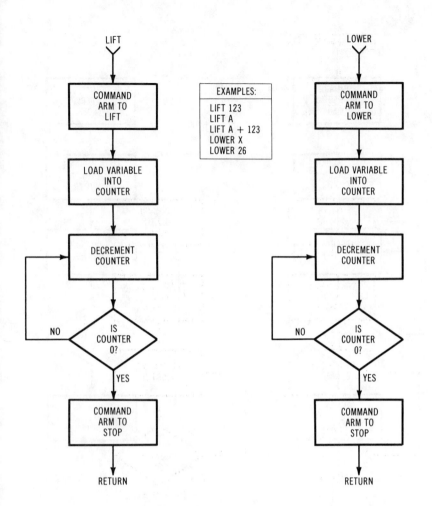

Fig. 6-9. LIFT and LOWER logic.

able after COMPARE must be added to determine which image you are comparing the value, retrieved during the SEE command, with. It may be possible to have many images stored in the robot's memory. The number of images is only determined by the amount of memory available. A 32 × 32 byte array might be somewhat large depending on the sensor used. In your system or any other system, you may find that a 16-bit × 16-bit or 32-bit × 32-bit array may be sufficient. In this case, you might be able to store many images. The COMPARE command would then select the image that you requested and compare it, bit by bit, to that image that was acquired through the SEE command.

If you've ever worked with images before, you will realize, quickly, that an actual bit-by-bit comparison probably will never pan out for the positive direction. It is almost impossible, given the correct lighting arrangements and any motion that the robot may have during the SEE phase, to be able to come up with an exact match. That is where the ROTATE command comes in. ROTATE, as shown in Fig. 6-13, will allow you to manipulate the image acquired during the SEE command, bit by bit, in any one of eight directions. To realize how this is done, looking at the figure, you can see that the image, stored in memory from the SEE command, is moved 1 bit in one of the specified eight compass point directions. This will effectively rotate the image slightly which might possibly help to match up the two images.

Image processing is a very complex field. It is not expected that, at this point in our discussions, you will be able to devise a ROTATE, SEE, and COMPARE system that will work 100% of the time. In fact, the COMPARE statement must have, embedded in it, a system for finding approximations instead of exact, bit-for-bit comparisons. It is, however, possible to imagine commands such as this, and it is important that you experiment and program vision-type commands to gain the experience and somehow augment the knowledge already known in robotics control. Who knows, you might discover something that no one else has. It's that type of a field, and it's this challenge that I hand to you.

Now we get down to the movement particulars of a robot in the ROBOL language. The next two commands, as shown in Fig. 6-14, should resemble something very similar to that of the LOGO language. I have shown an implementation using FORWARD and REVERSE in the first volume of this series. This was used with the Milton Bradley Big Trak command rotating base. Here it is very similar, a variable, in this case 1 to 99, showing steps that the robot might take in either the forward or reverse direction.

Taking this one step further, the RIGHT and LEFT com-

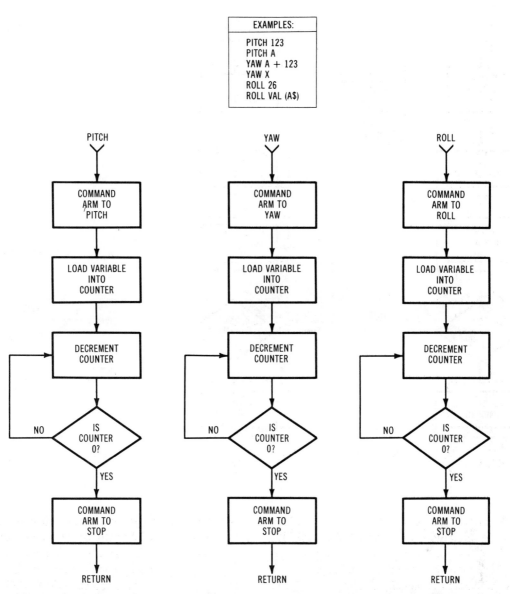

Fig. 6-10. PITCH, YAW, and ROLL logic.

mands, in Fig. 6-15, show compass points of degrees of rotation that the robot is to turn to. Now, in order to implement a system such as this would require a very sophisticated feedback mechanism. It doesn't matter, however, at this level what these compass points represent. They could represent actual magnetic compass points or simply location points around the axis of the robot. There does not have to be a full 360 of them. In the case of the Milton Bradley Big Trak there are only 60. Again, your judgment in the needs of the future should be allowed. It's possible that if you set up the system to allow 360° of rotation, then later on, as sensing techniques improve, you can utilize the fullness of the language without enhancing it again. Realizing, throughout this discussion, that ROBOL is being designed as a standard robot computer language that should not have to be enhanced to utilize various hardware technologies that may come and go. If you will notice, we've about run out of statements, as I presented them earlier in the list of ROBOL statements. This does not mean that a superset of this language could not be developed. For that matter, a subset, maybe known as Tiny ROBOL, may also be devised depending on your robot system. This might be the case if, per chance, the rover designed did not have a manipulator or had a different type of manipulator with a gripper that did not truly represent an open or closed physical property. You may find that developing a language such as this is very enriching, and I'm sure that to come up with a superset of these commands is a very easy task.

SOFTWARE DISTRIBUTION

You will notice that throughout the description of the ROBOL language there was very little discussion regarding the saving or retrieving of programs within the robot system.

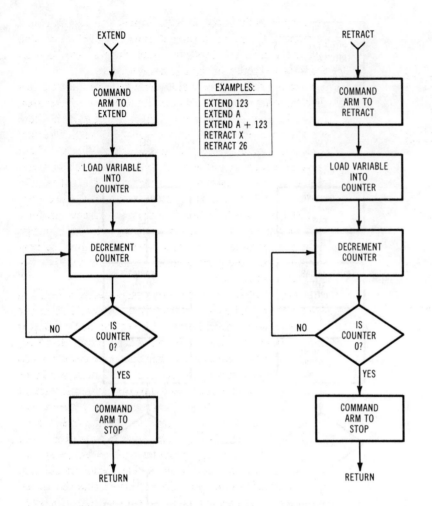

Fig. 6-11. EXTEND and RETRACT logic.

This may be known as storage. Storage in a robot might be done in read/write memory or it could be done more conventionally on disks or disk paks. This has yet to be determined in the personal computer field, as hardware advances have not allowed for low power storage mechanisms to be included as on-board entities. The HERO system allows for cassette tape I/O as does the RB5X robot. Other robots, being developed by major corporations, may utilize the minifloppy disk-type I/O that we all know as a basic part of every personal computer system. However, to have an on-board storage capability would enhance the robot all-in-one package. Distribution of robot software, to date, usually comes down to the deliverance of a system where the robot language such as the HERO executive is contained in firmware on read-only memory ICs. In the case of the RB5X robot, it is partially built into the microcomputer and the application routines are also in ROM. A way of devising a future plan for software distribution or updates to robots could be designed. Let's look at the use of some of the newer technologies in this application.

Semiconductor memories have advanced to where you can now buy an 8K × 8K static CMOS memory. CMOS is a nice choice for software distribution in that it is very low power and could be built into a small package with an integral battery that would allow the storage of robot commands without the necessity for external power. Therefore, units such as these could be shipped through the mail or hand carried between robots. Of course, semiconductor memories, with their fragile legs, sometimes become a problem with neophyte-type users plugging them in. If you are not familiar with plugging in ICs, then utilizing a 24-pin dip-type IC as a software carrier could become dangerous. There are new sockets and socket techniques that allow these to be plugged in fairly easily, however, they do not incorporate a battery holder. I've devised an idea called the data pak, which allows the use of CMOS memory and a small lithium battery built into a small microcassette-sized case with a standard type connector header on one side. A diagram of this type of a unit is shown in Fig. 6-16. Programs could be stored, up to 8K bytes in length, and disseminated among the masses utilizing this type of technology. The problem in using semiconductor memory in this application is that it is very easy to damage the part. CMOS memories are very prone to static damage and because there isn't much room inside of the data pak case, as you can see in the figure, for any type of buffering to be utilized, the memory might easily

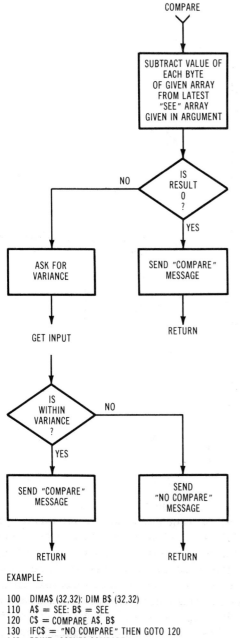

EXAMPLE:

```
100  DIM A$ (32,32): DIM B$ (32,32)
110  A$ = SEE: B$ = SEE
120  C$ = COMPARE A$, B$
130  IF C$ = "NO COMPARE" THEN GOTO 120
140  PRINT "SCENES COMPARE"
```

Fig. 6-12. COMPARE logic.

securing it to the robot. It requires that you disassemble a portion of the robot to plug it onto the CPU card. You actually handle the legs with your fingers. This can be disastrous. There must be another way.

Carrying the semiconductor memory thing one step further, there are new RAMs that are erasable through electrical pulses instead of ultraviolet light like the EPROMs. These are known as electrically erasable PROMS or EEPROMs. They come in various sizes all the way up to 8K also. These parts, however, are just as susceptible to damage and would require a special package as is the case of the data pak, however, they would not require a battery for back up. EEPROMs you will see used, later on in this chapter, for a very special type of system. I'm not going to let you in on that right now because I think that when you realize the possibilities that these parts give to a system, it will be very enjoyable at that time.

So far, we have been discussing the media involved in disseminating software from supplier to user. I have not mentioned floppy disks because I don't feel that the floppy disk of today is a viable entity on-board a robot. However, the newer microfloppies or 3½ inch disk drive assemblies do present a more compact arrangement and a more rigid disk arrangement than the previous floppy disk systems. It might be possible to allow robot software to exist in this form. It's something that must be experimented with. It has been found, in the personal computer industry, that disks are a viable way of allowing neophytes to load programs into computers. But this discussion of the media should also involve distribution. How do you determine the actual application that a person might have for a personal robot?

You cannot compare personal robots to personal computers. Personal computers may sit on a desk, table top or bench, they may be used for household management, financial management, or games. I don't foresee personal robots being used for household management or personal management. They might be utilized in some type of game; however, any motion involved would require a specialized set of circumstances. For instance, if you were to require that your

be destroyed by a human's touch on the connector pins. Also, it is possible for a nonexperienced individual to accidentally erase the memory because there is no write-protection. A form of protection on the write line could be used by physically disabling that input to the data pak after it was programmed at the factory. However, in this case it would be very easy to use a standard erasable programmable memory, or EPROM, to be used as the carrier for software. At this time, you can purchase software on EPROM or ROM from Heathkit and other type sources. Heathkit sells a demonstration ROM that, although is packed full of features, would require you to have special handling techniques used in

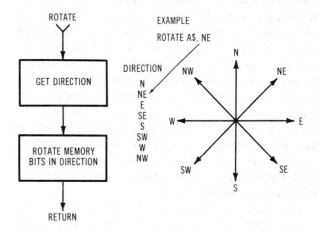

Fig. 6-13. ROTATE logic.

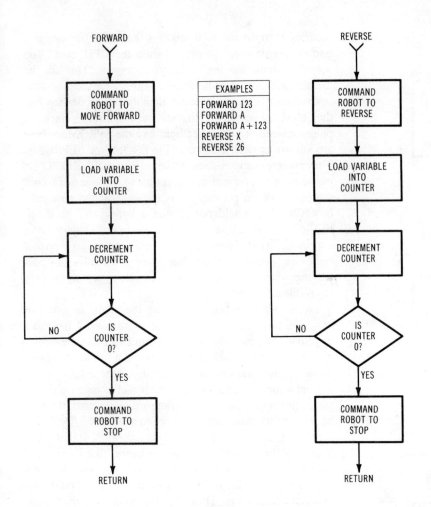

Fig. 6-14. FORWARD and REVERSE logic.

personal robot roam your house during the evening looking for intruders, it would require a specialized set of commands telling it the floor plan of your house. Earlier attempts by companies to view the future of the home robot have included the ability of the manufacturer to send a highly trained technician to your house to work on a floor plan for the robot. This is unacceptable, not only from a security standpoint, as there are many dishonest people, some may even become highly trained technicians, but it is also economically impossible. No company could afford to send a technician to everyone's house. Therefore, some type of self-programming must be allowed within the system. If you recall, Heath has made an attempt at allowing a system such as this with their learn mode within the HERO system. This is a very powerful construct. It allows the computer to not only take down the actions that you drive it through with a hand-held teaching pendant, but also turns them into a microprocessor language or actually acts as a minicompiler from physical action to assembly language. To be able to implement a system with very good accuracy would be a complex job.

Therefore, companies developing personal robots or robots for home use that might require specialized motion programming should expect that the distribution of their software would be limited to nonmotion items or each piece of software that required motion commands would have a built-in autoprogram mode such as the Heath learn mode. There is no question that distribution of games for use with personal robots will be widespread. There are many word games and learning games utilizing the voice output and input capabilities of these robots that would enhance a child's learning or an adult's entertainment. In fact, you may see this type of application become more widespread than any other; however, there has to be serious work on applications done before the masses will accept personal robots as easily as they have accepted personal computers.

HERO EXECUTIVE SYSTEM

While we are on the subject of HERO, having just gone over the learn mode, now is a good time to get into the Executive software utilized within the HERO system. This will not be a subroutine-by-subroutine description or anything as complex as that. The *HERO* book that I mentioned in the last chapter, published by Sams, goes into this type of detail. The purpose of stating the Executive system's architecture here is to show the systems outlook the designers used in developing the software. From the discussions of the

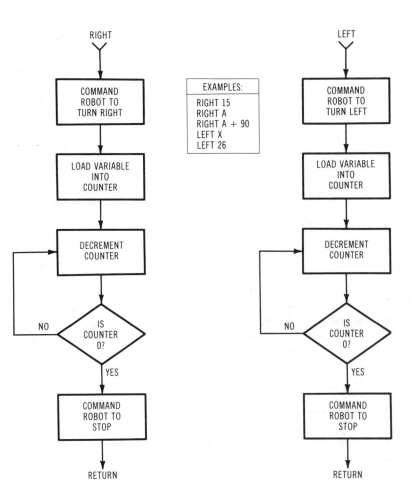

Fig. 6-15. RIGHT and LEFT logic.

HERO Executive system, we will move into a similar type of Executive system designed for the core processor personal robot that was developed in the past chapter. This is also an overview discussion.

Fig. 6-17 is a system block diagram of the HERO software architecture. As you can see, from a power-on or reset condition the software enters the Executive routine. This Executive is an overall running operation system that allows the operator to select, through the use of the keyboard, various modes of operation. If you will notice off to the side of the beginning of the Executive routine, there is a dotted line area from an external Executive. When the reset condition is finalized the 6808 microprocessor on the CPU card of the HERO system looks at a location in memory (2000 hex) for a certain code. If the data located at this location matches the code that it is expecting to see, it assumes that an alternate Executive program is to be run. This gives users the possibility of plugging in or somehow running their own operating system or an operating system supplied either by Heath or some other vender. On the CPU card is an extra ROM socket that may be configured for any industry standard pinout up through an 8K × 8K part. The location of this ROM socket is selectable and may be selected for this address. Therefore, even before the Executive routine is run, Heath has allowed for a completely new system. This forethought will help them in the future and shows the overall systems thinking that went into the product.

When you enter the Executive routine, the robot recites the word "ready" through its speech option and shows the name HERO and the version of the firmware included in the robot on the six-digit LED display. From here, it goes through a normal initialization routine of certain variables. Then the keyboard is constantly strobed looking for input that would be in the form of operator commands as to how to proceed. It is from here that you have four branches or roads that you can take. As seen in the diagram, it is possible to select the program mode, which will initiate the running of machine-language routines, and the robot mode, which would allow both machine language and embedded higher-level robot commands to be run. Utilities may be called, which we went over briefly in the last chapter, for setting the real-time clock calendar and loading or saving data to cassette tape, or the debugger, as mentioned earlier, may be entered.

The most complex of these four modes is the robot mode. Fig. 6-18 is an overall systems diagram of how this mode operates. As you can see, when you enter the robot mode, a program is run much like a native language 6808 code program would be processed; however, each op code byte

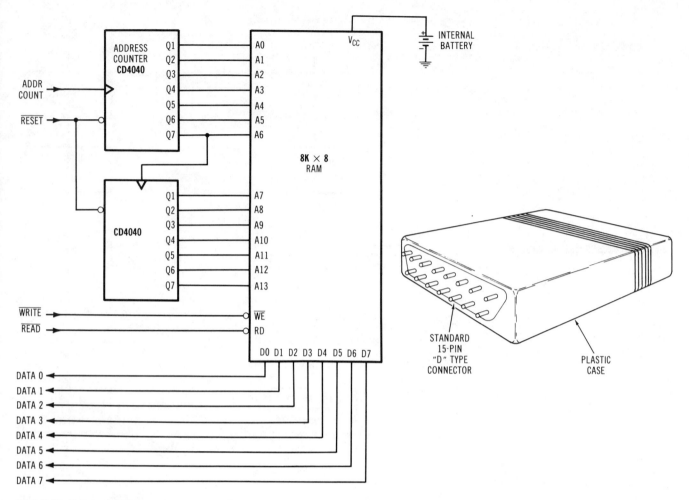

FIG. 6-16. Data pak idea.

that is brought into the microprocessor is checked against a table of valid 6808 native language op codes. If it is found that the command interpreted is not a 6808 op code, another table is then consulted to determine if it is indeed a valid robot mode or robot command op code. Once again, Heath has provided hooks that allow it to expand the operation system without digging into the ROM. Here, when the robot mode table does not pan out, the program will jump to a location in lower memory where it is expecting to find another jump command to an enhanced robot command table that you may provide or another vender may provide for you in the form of external add-on memory. During power-up, this place in lower memory is initialized with a return to subroutine, which will automatically bring it back to the robot interpreter. Assuming that you have not changed this memory location, that will aways be the case. However, after the "ready" prompt, it is possible for the user to go into the debugger and physically change the memory locations that are used as pointers throughout the lower RAM area. As we go on in the discussions, you will see that there are several places that Heath allows you to expand its system. Keep in mind the abilities of this initial robot offering and how they have designed their Executive system to incorporate any type of add-on capability.

Moving along, let's assume that a robot command is valid. It then will jump to the area in memory called out for that particular subroutine to be utilized. What type of routines are there? Table 6-1 is a list of the various robot commands supported, at this time, for HERO 1. As you can see from the table, there are 38 varieties of commands. Many of them deal with the motors used in the arm, head, and base. When the enable commands are run for each of the senses, realize that it is then that the senses are turned on and power is applied to them. Simply enabling them will not read data from them. The disable commands for the senses, conversely, will turn off the power to these boards. At this time, let's go into the MOTOR MOVE commands to see how they are implemented in the HERO system.

Table 6-2 is a breakdown of the syntax used for the MOVE commands. There are several varieties of MOVE commands, as can be seen in the table. Each one supports a different approach to programming. As you can see, there are several immediate-type commands. The immediate variety allows for the specifics relating to motor type, direction,

Table 6-1. HERO Robot Commands

ABORT drive motor	DISABLE motion detector
ABORT steering motor	DISABLE display
ABORT arm motors	SPEAK continue (indexed)
ABORT speech	SPEAK wait (indexed)
BRANCH if base busy	SPEAK continue (extended)
BRANCH if steering busy	SPEAK wait (extended)
BRANCH if arm busy	CHANGE to machine language
BRANCH if speech busy	SLEEP (immediate)
ZERO	PAUSE (immediate)
RETURN to executive	JUMP if speaking (extended)
CHANGE to robot language	MOTOR MOVE wait absolute (immediate)
ENABLE light detector	MOTOR MOVE continue absolute
ENABLE sound detector	MOTOR MOVE wait relative
ENABLE ultrasonic ranging	MOTOR MOVE continue relative
ENABLE motion detector	MOTOR MOVE wait absolute (indexed)
ENABLE display	MOTOR MOVE continue absolute (indexed)
DISABLE light detector	MOTOR MOVE wait absolute (extended)
DISABLE sound detector	MOTOR MOVE continue absolute (extended)
DISABLE ultrasonic ranging	MOTORS MOVE ALL absolute (immediate)

and speed to be supplied as 2 bytes following the command byte. Looking at one of the immediates, the MOVE MOTOR wait absolute immediate command, denoted by op code C3, we see that the following 2 bytes represent the motor to be selected, the speed and the direction. In this particular case, the wait absolute applies to the fact that the program will not stop executing until this motor command is complete. Absolute specifies that the position dictated in the last byte and the first 2 bits of the second to the last byte is an actual location as opposed to a relative location that would be commanded through the use of a D3 op code.

As you can see from the table, it is possible with one of these commands to specify any of the eight different motors in the robot. Relative commands allow the microprocessor to determine the motor position by counting a number of positions from the present position. Executing a MOVE continue absolute or relative will allow the machine-language programs to still be executed while the particular selected motor is moving. In this way, motor locations come back as interrupts to the processor and it is transparent to the programmer that the move operation is still in process. This allows for faster operating programs. The last MOTOR MOVE command or the MOVE ALL, specified by code FD, requires 7 motor position bytes that give absolute position for the motors in order of extend, shoulder, rotate, pivot, gripper, head, and steering. Each motor, however, will be operated one at a time as the program steps through. Looking further at the table, we see that there are two 2-byte varieties of the MOTOR MOVE command. These allow for the use of the index register to point to a location in memory where the 2 motor control bytes are stored for that motor. This allows the programmer to write a program that might possibly modify those bytes according to the operation of the program. This is another example of how the Heath Co. has been able to think ahead to the use of such a system and allow as many options

Table 6-2. Motor Commands

IMMEDIATE COMMANDS	
C3 SS XX	MOTOR MOVE, wait absolute
CC SS XX	MOTOR MOVE, continue absolute
D3 SS XX	MOTOR MOVE, wait relative
DC SS XX	MOTOR MOVE, continue relative
FD *	MOTORS MOVE ALL, absolute
INDEXED COMMANDS	
E3 OO	MOTOR MOVE, wait absolute
EC OO	MOTOR MOVE, continue absolute
EXTENDED COMMANDS	
F3 MM MM	MOTOR MOVE, wait absolute
FC MM MM	MOTOR MOVE, continue absolute

*Seven motor position bytes: extend, shoulder, rotate, pivot, gripper, head, steering.

XX = distance, position
SS = select motor, speed, direction
OO = offset number from index
MM = memory address of data

as possible for the programmer.

Looking back at the list of HERO commands in Table 6-1, we see that there are four ABORT commands that are 1 byte in length. These will allow the program to immediately stop the actions of the drive, steering, arm motors, or the speech output. These may come in handy, especially in the motor movement area, in case the possibility of collision exists or if some sort of external contact sensor has detected an object in its path. Also, as you can see in the list, there are four BRANCH instructions that allow the programmer to institute some logical decisions based on whether the base motors are busy, the steering motors are busy, if the arm motors are busy, or if the speech accessory is busy. In the case of the speech, you may have a program that outputs constantly to the speech accessory, and it would go through a BRANCH if speech busy, waiting for the speech accessory to become idle before sending the next set of phoneme codes.

Table 6-3 outlines the speaking commands. There are four of them that allow the string of phoneme codes to either exist at a called-out memory location, which is denoted by the SPEAK continue and SPEAK wait extended instructions, or to be pointed to by the current contents of the index register.

As is the case in the motor routines, SPEAK continue will allow the voice synthesizer to be speaking the phoneme codes while the program goes on looking for the next robot command. SPEAK wait will stop all actions until the speech accessory is again idle. To complete the information on the speech synthesizer, the Votrax unit, as I said in the last chapter, utilizes phoneme codes or portions of sound that make up words. We saw a table of allophone codes in the last chapter and here, in Table 6-4, is a full list of phoneme codes for the SC01A synthesizer chip. It is these codes that would be placed in memory, pointed to by either the extended 2 memory bytes after the SPEAK command or pointed to by the index register. The speech processor would then pull

Table 6-3. SPEECH COMMANDS

INDEXED COMMANDS	
61 OO	SPEAK, continue
62 OO	SPEAK, wait

EXTENDED COMMANDS	
71 MM MM	SPEAK, continue
72 MM MM	SPEAK, continue

OO = offset number from index
MM = Memory address of phoneme string

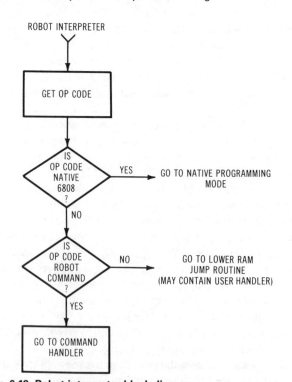

Fig. 6-17. System block diagram of HERO software.

Fig. 6-18. Robot interpreter block diagram.

Table 6-4. Votrax SC01A Phoneme Codes

Phoneme Code	Phoneme Symbol	Duration (ms)	Example Word	Phoneme Code	Phoneme Symbol	Duration (ms)	Example Word
00	EH3	59	jackEt	20	A	185	mAid
01	EH2	71	Enlist	21	AY	65	mAid
02	EH1	121	hEAvy	22	Y1	80	Yard
03	PA0	47	-PAUSE-	23	UH3	47	missIOn
04	DT	47	buTTer	24	AH	250	gOt
05	A2	71	enAble	25	P	103	Past
06	A1	103	mAde	26	O	185	mOre
07	ZH	90	meaSure	27	I	185	pIn
08	AH2	71	hOnest	28	U	185	tUne
09	I3	55	inhibIt	29	Y	103	anY
0A	I2	80	Inhibit	2A	T	71	Tap
0B	I1	121	inhIbit	2B	R	90	Red
0C	M	103	Mat	2C	E	185	mEEt
0D	N	80	suN	2D	W	80	Win
0E	B	71	Bag	2E	AE	185	dAd
0F	V	71	Van	2F	AE1	103	After
10	CH*	71	CHip	30	AW2	90	sAlty
11	SH*	121	SHop	31	UH2	71	trAdition
12	Z	71	Zoo	32	UH1	103	About
13	AW1	146	AWful	33	UH	185	cUp
14	NG	121	thiNG	34	O2	80	fOr
15	AH1	146	fAther	35	O1	121	abOArd
16	OO1	103	lOOking	36	IU	59	yOU
17	OO	185	bOOk	37	U1	90	yOU
18	L	103	Land	38	THV	80	THe
19	K	80	Kitten	39	TH	71	THing
1A	J	47	JuDGe	3A	ER	146	bIRd
1B	H	71	Hello	3B	EH	185	gEt
1C	G	71	Get	3C	E1	121	bEfore
1D	F	103	Fast	3D	AW	250	cAll
1E	D	55	paiD	3E	PA1	185	-PAUSE-
1F	S	90	paSS	3F	STOP	47	-STOP-

*/T/ phoneme must precede /CH/ to produce "ch" sound.
*/D/ phoneme must precede /J/ to produce "j" sound.

Table 6-5. Typical HERO Program Examples

```
           LIGHT DETECTION

     3F         ENABLE robot language
     41         ENABLE light detector
     20 FE      BRANCH here continuously

     3F         ENABLE robot language
     45         ENABLE sonic ranger
     83         CHANGE to machine language
GET  96 11      GET # of sonar hits into ACCA
     BD F6 4E   JUMP to reset display subroutine
     BD F7 AD   JUMP to display ACCA on display sub.
     CE 20 00   LOAD index reg with value 2000
LOOP 09         DECREMENT index register
     26 FD      If index is not zero go to LOOP
     20 F0      When time is up get another from GET
```

these codes out, sequentially, speaking each one until it received an FF code. At that time that would denote the end of a spoken sequence of commands. The robot language interpreter would then go back and look for the next command in the series.

Some brief examples of the mixture between robot language and machine language are shown in Table 6-5. The use of the 3F command, which, in 6808 language, is the software interrupt, turns the robot interpreter on. As you can see before the 3F command is utilized, the robot is running machine language. The code 83 allows you to switch back from robot language to machine language. Wherever speed may be a problem, that is speed of execution, it's a good idea to switch to machine language, providing you don't have any robot commands embedded in that code. The 6808, upon receipt of a robot code, will crash the program; however, if the robot interpreter is on line, any command will be processed normally. The penalty you pay for the look-up through the two tables, mentioned earlier, is a tremendous reduction in operation speed. As you can see, all possibilities have been accounted for in the design of this robot language.

Now that we have a general overview of the HERO Executive system and how it relates to the robot interpreter, the utilities that were mentioned in the last chapter, and the debugger, you have a good feel for one way of programming a robot. This will allow a neophyte programmer or even the most advanced engineer to be able to write sophisticated applications programs using this robot language. However, the nonprogrammer, or the uninitiated electronics entertainment buyer, upon reception of a HERO unit, would not be able to program at this level. In fact, it is doubtful that the person would be able to comprehend the idea of the interpreter commands very quickly; therefore, as I have said, Heath has built into the robot a self-learning routine. This is called the learn mode, and it is entered from the utility phase.

LEARN MODE

Automatic programming has always held a certain fascination for me. It seems obvious that the way of the future will utilize systems that employ automatic programming languages. Therefore, it would not be necessary to be versed in any one type of machine. All you have to do is the physical motions or run the machine through its paces and then, it runs itself. This is easier said than done. Heath Co., however, has been able to accomplish this through the use of the teaching pendant. In industrial robots, teaching pendants have been used for years. They allow an operator to step a robot manipulator through its paces, step by step, as it learns each arm position. It learns these in a series of stops. The HERO robot, however, builds a machine-language-like set of codes each time the pendant is used in the learn mode. For instance, if you will remember our discussions of the MOTOR MOVE absolute immediate commands, this same command set and syntax is automatically generated by the learn mode when you place the teaching pendant in the forward or reverse direction and the controls to operate the robot. When the robot stops at a given point, the command is finished and compiled to look like a standard MOTOR MOVE command. Let me go over this more slowly and in detail to show the process that goes on. The teaching pendant has a series of switches built into it that allow one to drive the robot. A close-up representation of the pendant operator panel is shown in Fig. 6-19. It is arranged as a gun-type arrangement with a trigger on the bottom that the index finger fits into nicely. The thumb is used to press the rocker switch in the lower middle of the control panel, which selects from two different types of commands, depending on the master rotary switch shown in the center position.

Placing the rotary switch in the head position, which is to the far left, the function slide switch in the arm position,

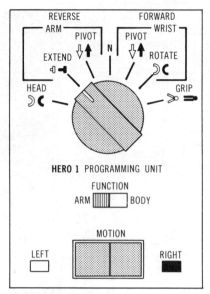

(A) Control panel.

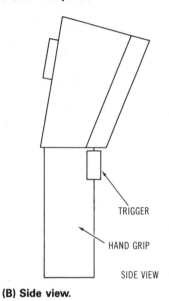

(B) Side view.

Fig. 6-19. Teaching pendant control.

which is to the left, and pushing the trigger on the bottom of the unit will move HERO forward at a high rate of speed. At this time, in the learn mode, the op code C3 has been placed in a memory location. You will notice on the display the units of motion incrementing as you move the drive motor. When you stop depressing the trigger, the entire command, made up of three bytes that represent the movement from the place where you started the command to the place you are now, is positioned in those memory locations. The next time you push the trigger another command is then put behind the first command. As you can see, this will build up a repertoire of commands, in sequence, which might be called a program. It is possible, through the keypad, to be able to step backwards through this program and edit it just as you would any program that you entered yourself but, through the use of the

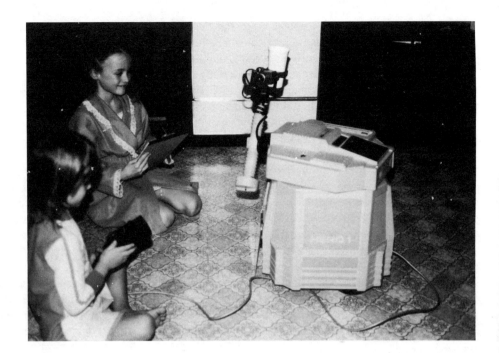

Fig. 6-20. Photograph of small child teaching HERO.

teaching pendant, anyone can program the robot without having any prior knowledge of any electronics or programming language. Fig. 6-20 shows a small child teaching HERO a series of commands without the use of any prior knowledge.

This last figure illustrates the power of this command. If you'll remember from our discussions of the LOGO language, children were set down in front of computer screens and taught the basics of geometry by moving around a small diamond on the screen, which represented a rover robot. Here, children may sit down and plan out the motion of an actual robot and, without having to use a typewriter style keyboard or even having to remember the words forward, back, right, or left, may command that robot through a series of programming commands and build programs. At that time, when the program is finished and operates the way the child wants it to, he or she may then go back through the computer's memory and write down the steps that it took to accomplish that task. Of course, here is where some adult-level supervision must enter the picture, because what comes out of HERO are machine code-like commands even as robot commands. They are not as easily understood as forward, back, etc.; however, through the use of some simple programming and an external terminal connected to HERO's I/O port, it might be possible to convert these robot language command codes into English language statements. This would be something to experiment with in the future, if you are interested in further experimentation with a HERO-type robot. At this time let's move on to our advanced core processor personal robot design that was developed in the last chapter. Let's specify a language, like the HERO language, that allows us to control that robot. Let's also work on some specific I/O device drivers and pay particular attention to the human interface.

ADVANCED ROBOT LANGUAGE

If you will recall, in the last chapter, we built a new personal robot based around the distributed processing techniques learned in Chapters 1 and 2. This distributed processor utilized the SCU20 serial I/O command set IC to perform all the I/O functions in a personal robot. I repeat here the functional block diagram of what is known as the core processor personal robot in Fig. 6-21. Let's examine an overall executive software system that might interplay with the hardware developed then. Recalling the structure of the HERO system, after the "ready" sign-on message you had the possibility of keyboard command in any one of four areas. Before that, an automatic jump to another alternate operating system happened. In that case, a different set of commands could have supplanted the keyboard operated ones. In our system, let's try and go one step further. You may find that may be hard to do, but looking at the possible applications for a personal robot and forgetting about the engineering application or engineering users that might want to dabble into the intricacies of the hardware and software, let's try to gear this robot toward the nonengineer but still allow a great deal of intervention within the system.

In order to do that, you might want the robot to power-up and basically be at the level of possibly a two- or three-year old child. You'll want it to understand things such as self-preservation, possibly how to get around, and general needs such as these. It would be nice to be able to program the robot as you would a child. Each thing it learns it would then put away for future reference, never forgetting anything, unlike a child or human being. Of course, this wish list may be hard to achieve, depending on the amount of memory involved. Our core processor, if you recall, only showed an 8K memory on the CPU board. This does not mean that we couldn't change the memory arrangement or add to the processor as time goes on with some sort of mass storage system such as disk, tape, bubble memory, or maybe some semiconductor memory that hasn't been developed yet. Throughout our discussions, please keep in mind this ability to teach the computer or the robot everything.

Right from power-on you'll want the robot to be interactive. The human interaction capabilities of the core processor are very important. The ready message of HERO is very good in knowing that the machine has powered-up, however, it doesn't tell you much. It may be possible, and maybe even desirable, to use the word "ready" for the core processor. However, from there the robot would not exclusively look for a keypad entry. Let me suggest an approach that you may use after power-up. During this approach it will be necessary to have some nonvolatile memory within the system. This would possibly mean that the 8K of RAM should be battery backed up so that it does not go away when the power is cycled, as is the case in the HERO robot. The reason I need nonvolatile memory is that I would like the robot to power-up and look for the status of the real-time clock calendar outputs. There may be a byte in memory that tells me whether they have been set previously or not. If they have not been set, the first thing the robot should ask the operator who turned it on to do is to set the correct time. By asking this there are a number of ways of prompting the user for this input.

Human interaction should be as natural as possible. Asking an operator to set the time should not be mechanical in nature. It should not be "enter time" or "enter date." There should be a reasonable interaction on the human personal level between the robot and the operator. You want a feel for the fact that this is a friend, this is an almost living entity. You might want it to announce its name and then ask you to "Please, for your own benefit, give me the correct time and date so that I may be able to serve you better by utilizing this information in future commands." At that point, you've put the user at ease and you've given the user the ability to be in command. Entering in the time and the date should also be as natural as possible. For instance, not having to wait for leading zeros on single-digit hours, such as 6:00 in the morning, should not have to be entered as 06:00. You should allow a variable syntax when operating with a human. It takes a few more bytes of code to do this but, in the long run, will provide a superior product. Therefore, we start to build our software functional diagram for the core processor. The beginnings of it are in Fig. 6-22.

From the end point in that figure, where do we go? HERO allows you to go into the program mode, robot mode, utility mode, or debugger. Utilities are nice things; they allow the operator to interact through the keyboard and do specific known things with the robot, such as load a new program, dump an older program, or set the time. We've already allowed the setting of the time, and it should not have to be done again, unless you do want to be able to change the time, such as in the case of time-zone changes caused by robots that may be moved to different areas of the country or,

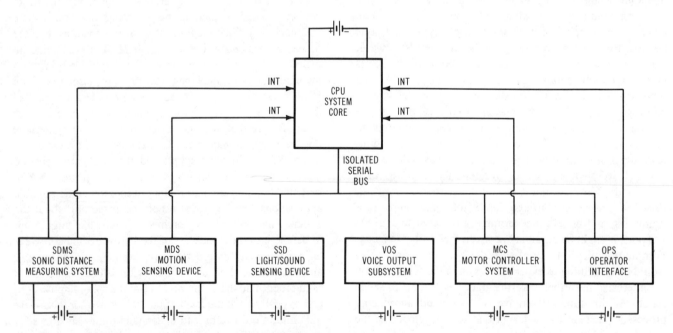

Fig. 6-21. Functional block diagram of core processor personal robot.

maybe, daylight savings time changes. In the case of daylight savings time and even time-zone changes, it might be nice to have a higher-level utility that will allow the robot to take care of the hunting for the correct hour. Therefore, a utility mode for time changes might look like that shown in Fig. 6-23. This is much more sophisticated than the HERO variety of robot software. The robot will ask you, upon entering this mode, whether you are changing from daylight savings to regular time or if you've changed time zones. If you have changed time zones, it then asks you which time zone you are in. This takes the programming and the command set knowledge out or away from the operator and allows the robot to prompt you or bring you along, for it is you who must conform to the robot, not the other way around. It should be the robot making it easier for you. After all, it is supposed to be a tool to help you or to somehow enhance your life or your entertainment.

The utility in Fig. 6-23, however, does not show how it is called. This may be somewhat of a dilemma. How do you get to tell it that you want to change the time without having to memorize certain keypresses on the keyboard? For that matter, how do you start it running a program or how do you enter programs? Fig. 6-24 is my first stab at a higher-level command selection routine. It's not that easy to command a robot through a keyboard, without memorizing codes. You could, however, arrange the keyboard with legends that denote certain command modes. These legends, shown in Fig. 6-24, allow the operator, through a quick glance, to select various operating conditions. Now, let's look at the keyboard in that figure. You can see that the legend by the C button will allow you to change the time. Also, storage options are allowed by pressing D. Let's stop right here and think about storage for a while. In HERO, when you depress the utility mode, button 3, and then you depress another key, denoting either cassette dump or cassette load, you are prompted for a starting address. This starting address is a place in memory where the program exists or is to exist. This assumes that you know that there are memory locations and that there is a program that requires memory locations. What I'm getting at is that you are a sophisticated user. The core

Fig. 6-22. Initial software block diagram.

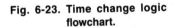

Fig. 6-23. Time change logic flowchart.

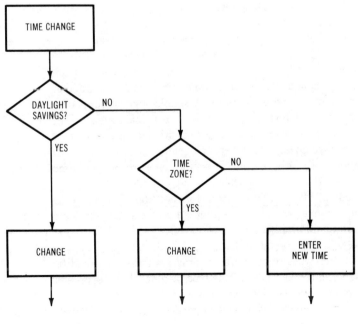

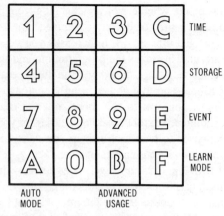

Fig. 6-24. Proposed operator interface keyboard.

processor is designed for nonsophisticated, nonengineering users. So, how would you specify a program? The use of names comes up first. However, names imply a keyboard with a full alphabet. Core processor initial design, as shown in the last chapter, only provided for a hex keypad. Names are out. Numbers aren't too bad, especially if you can allow more than one program to be resident in memory at any one time. Let's say four programs. How big are they? Well, that depends on the amount of memory that exists. But it shouldn't depend on how large the programs are.

Programs could be stored in a variable format. It may be stated, in the core processor's users manual, that up to four references for programs may be used; however, a certain limit on the size of any one program would exist. You could, of course, allow one program to take up the entire memory. This should be allowed. Four references tell the programmer who is building the executive system, which presumably is you, that you should allow for syntax, after the storage option of which program is being loaded into memory or which program is being operated or dumped from memory. Let's take that one step further. The E key is a special mode of operation. This will allow you to enter in events. Events being times with certain dates that certain programs or portions of programs should be run. This is probably the most useful a robot could be. For instance, let's say you program him to announce, in the morning, that you have a doctor appointment. Sure this is a very simple application, people have been touting this application for personal computers for years; but to have a robot be able to drive to the side of your bed, take down your covers and gently nudge you and then remind you that you have a doctor appointment some time during the day, is much more effective than getting up, loading in the operation system for your Apple® or other type of computer and having it print out on the screen the same information. There is a certain immediacy to the ability of the robot to get you out of bed.

Other types of programs or program messages could be done on a timed basis. This type of timed event operating method could become very powerful. You must allow it the freedom to be that powerful. A general skeleton structure of that type of a command or executive routine for the E or EVENT utility is shown in Fig. 6-25.

From here, we're going to skip over the F command. That's something we'll go over in the artificial intelligence section of this chapter. Move to the lower left-hand corner of the keypad to the A button. You'll notice that the legend around there is called auto. You might guess that I enjoyed the HERO automatic programming LEARN MODE enough to place it into the core processor personal robot. That's exactly it! To be able to build your own programs by driving the robot around by hand is very powerful. I would like to preserve that in this robot. Therefore, the auto mode here works almost exactly like the HERO LEARN MODE. There is one hitch, however. What I did not mention in the LEARN MODE for HERO is that when you enter it, you must enter the starting location in memory and the ending location in memory for the program you are building. This is, once again, assuming you are an engineering type of person. The core processor personal robot does not assume this and will

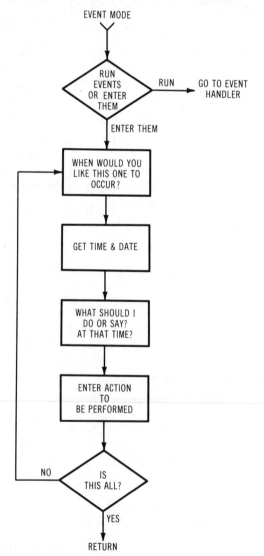

Fig. 6-25. Preliminary EVENT-mode structure.

not ask you for that. It will, however, ask you for the program number, if you wish to number it. This will help in storing and recalling it later. I don't feel that asking for a single-digit number from one to four is that restrictive. In fact, at this point, I can see a problem with only allowing the numbers one to four, remember that we would allow up to four programs to be resident within the robot system at any one time. This does not, however, mean that we have to restrict the number of the programs to between one and four. Say, for instance, you had 27 programs that you had amassed over a period of a month, with this robot. At any one time you may want to have program numbers 1, 13, 26, and 8 resident in memory at the same time. Restricting it to numbers one through four only allows you to develop four programs or forces you to rename your programs with the same name. That would promote confusion. The idea of this robot design is to eliminate confusion and enhance the use of a personal robot. Therefore, I suggest that we open it up and allow any number, maybe up to three digits, to specify a program.

Getting back to the automatic routine, entered by pressing A, you are prompted to ask for a number that you might want to call this program. This prompt and any other prompts in the system are verbal. It would be spoken to you, in whatever language you normally speak, and you would then either reply or not reply. If you did not reply to the number, the robot must assume that you are not interested in naming this program. It, however, would name it for you. It would name it with a number that has not been used, to date. This may be a little premature and really should be in the artificial intelligence section but it comes in handy, right at the moment, to talk about it. Every command that you enter into this personal robot should be remembered by the robot. If, up to now, you've only dealt with programs 1 through 18 and you've never numbered a program larger than that, entering into the auto mode and going immediately to driving the robot should make the robot make the decision for you and call this program number 19. However, when you finish driving the robot, and we'll get into how to drive this advanced core processor robot later, it should then ask if you want to call it 19. This gives you that little bit of control that is necessary. It may be that you just wanted to try something out or show something to a friend and did not want to use up a number or you have no intention of storing the program anyway. This gives you that out. It's just one more indication of human interaction in programming.

Now to how to drive this system. Driving HERO with the hand-held teaching pendant can sometimes become cumbersome. You've got to remember a series of positions of switches and every time I do it, the rocker switch, which denotes right or left, is in the wrong position. In the end, I end up having to either edit out a number of wrong way directions or have a robot that operates in a jerky fashion. Of course, the ultimate way of commanding the robot to do something would be to tell it "watch me," and then you perform the action. Well, at this stage of the game, that's not possible. It might be possible to outfit you with some sort of suit that would feed back information to the robot, as far as positions of your arm and movement of your legs, but this would be rather expensive, and most people would not don the suit for fear of electrocution anyway. Therefore, something to use with the hand in conjunction with actual motion should be devised.

This, you will find, may become the hardest part of determining how the new system should work. If you based it on a car-type driving wheel and gas pedals, you could then set up something at a kitchen table or desk that allows you to sit and watch the robot while you drive it; however, this doesn't give you the visibility, into its actions, when it turns the corner and goes into one of the bedrooms. You cannot see what the robot is doing; therefore, the pedal action and two-handed steering wheel must be out. Another method of driving could be done through the use of a joystick.

Joysticks are probably the most familiar control for non-computer-type users today. Almost everyone has operated a video game where a joystick moved a space commander through a field of asteroids, shooting up enemy war planes. Joysticks are fairly easy to control and they force your hand to point to the direction you actually want the controlled entity to move in. Using a hand-held joystick, control of a personal robot should be pretty easy. This will suffice for the base wheel drive; but how do you command a manipulator that might have six stepper motors in it, with a joystick?

This is not easy either. This requires you, the programmer, to devise a very intelligent I/O device driver. I want to use two joysticks to control this robot. One will drive the base as we just described. The other will move the manipulator and allow with a button on the top of the second joystick the ability to open and close the hand. But how do you get the arm into the position that you want it? Drive it there with the joystick. This means that all motors, in concert, must be able to react a variable amount depending on the distance or direction they are commanded. A panel showing a representation of this drive control is shown in Fig. 6-26. One of the

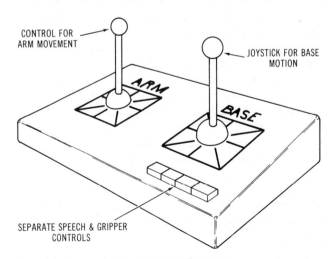

Fig. 6-26. Representation of AUTO MODE drive control panel.

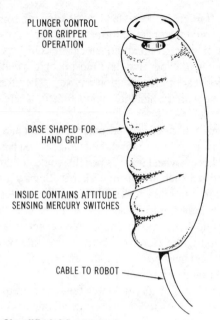

Fig. 6-27. Simplified drive control.

first objections you will see to this type of panel is the fact that it looks like a radio-controlled model airplane panel. There are too many controls on it and an unsophisticated user will not be able to command both hands simultaneously without making mistakes. This might be true; therefore, maybe you'll want to provide one joystick that can be utilized for either the arm or the base. Assume then that the operator would not be commanding the base at the same time as the arm.

This modified control panel now resembles that of Fig. 6-27. "Where did the box for the switches go?" you ask me. "All I see is a stick." That's right. This is a motion-controlled joystick. At the present time you can buy motion-controlled joysticks for use with video games. There is no base, and the tilting motion from the horizontal is what determines the distance you are commanding. Tilting the stick left, right, forward, up, and down will then move the base left, right, forward, or reverse. There should be a small switch, which can be activated, which would then signal the arm is to be commanded. A button on the top of the control stick would then serve to open and close the gripper. Holding the button down would result in the gripper opening slowly until you got it to the point where you want it. Letting go of the button would then close the gripper to where it could close up against the object. An automatic force-limiting arrangement should be built into the gripper that would not allow it to crush tiny objects. As you can see from the first attempt in Fig. 6-26, to this more sophisticated user-oriented control, shown in Fig. 6-27, there is a world of difference. The engineer, however, might go for the first approach. It allows a much greater degree of control and may be offered as an option.

Speaking of the engineer, we can't forget that sophisticated individual who will want to get inside the inner workings of this machine and enhance it even further. The B key on the keyboard allows you to get into the internal operations of the processor. It opens up a whole new world that a normal user will not get into. Depressing B will allow debuggers, utilities, and programming through the keypad. You might also allow the ability to plug into the serial bus through the use of an expansion SCU card that allows the engineer to hook up a standard ASCII terminal to the robot. Inside the robot executive an alphanumeric ASCII dialog would then allow the user to program in machine-language op codes instead of hex byte codes. These machine-language op codes are also known as mnemonics. They allow the user to use the words such as LOAD, IMMEDIATE, AND, ADD, etc., instead of two-digit commands. Having a built-in assembler/editor might also help in the microprocessor language of the robot. The sky is the limit on this command. As much as can be physically fit into that operating system ROM should be given there for the engineering user.

Throughout these discussions of the programming of the advanced core processor, you'll notice that the limitations in the original memory estimates have always been too limiting. Maybe it's time to go back into the processor card and rethink the memory arrangement. Newer memory-management integrated circuits are available that allow processors that can only address, directly, 65,000 bytes of information to address many megabytes of information through these memory-management techniques. It would be a good idea to not limit the size of the memory in this type of an advanced personal robot. Just speech commands can take up a lot of memory, and you'll want to add as many speech commands as possible. You'll notice, throughout this whole discussion of the advanced processor personal robot, specific device drivers have not been addressed. As mentioned in the last chapter, devices come and go. Voice recognizers today may not resemble the recognition system outputs of tomorrow. Take this into account. Allow for a design with modular I/O drivers that can be picked and placed depending on the requirements of the technologies of today. Keep some on the shelf for the technologies of tomorrow. Speaking of the technologies of tomorrow, let's look at incorporating artificial intelligence into a personal robot.

ARTIFICIAL INTELLIGENCE

By artificial intelligence I mean, the ability for the robot to somewhat think on its own. So far, it has given the appearance of a thinking robot by interactively conversing with you at an adult level. However, the learning capabilities, as I mentioned in the beginning, will prove to show the artificial intelligence of the robot. Teach the robot like a child, I said earlier. That's the best way to interact with a piece of electronic equipment as sophisticated as a robot. How do you teach it like a child? You will require some nonvolatile, changeable program memory. I alluded to this when we spoke about the EEPROMs earlier in this chapter. Right now, I would like to speak about a concept that I call the Lombardo data pak.

The name of this data pak, Lombardo, comes from its creator Joseph Lombardo, a friend of mine who came up with an idea, about a year ago, for a changeable, alterable program memory that may be used like Atari or video game cartridges plugged into a robot. Each cartridge may contain many applications. These applications can be anything that you might decide. If you are familiar with program generators for personal computers, think of the Lombardo pak as a built-in program generator memory card. To have these both on the same card will require a great amount of memory. Let's look at the operation of the Lombardo pak.

A simple programming technique, shown in Fig. 6-28, has the robot prompt the operator with a question. The answer to the question is a yes or no condition. The yes or no entered is then stored forever in memory. The next time the robot goes down this path, it will take the path selected by the user previously. This can be very powerful. For instance, if you are programming an application pak to become a security system, it might ask you, "What time should I turn on my security sensors?" You then would program in the time. It then may ask you to drive it around the route that it should follow. Take your control stick and drive it around. From there, it may ask you what to do when an intruder is found. Then, you would give it the option. It may then give you a series of possibilities that it could do when an intruder was detected. Select from these options. That may be the end of the programming sequence, as was shown in that figure. The next time that application is run, it will not ask you how to set up, it will just do it, and it will continue doing that until you, somehow, erase the memory in that applications pak. The Lombardo pak allows this ability to erase, but only through a certain programmed way. By "programmed way" I mean that you do not want just anybody erasing your hard-learned applications; therefore, password protection should be built into any application that is blown into a Lombardo pak. Where does this self-learning end? Actually, there is no limit to a self-learning computer, depending on how much memory you have. Let's look at another learning application.

The following shows a simple dialog between operator and advanced robot:

> *STAR**
> WHAT IS A STAR?
> *A STAR IS A SELF-LUMINOUS, SELF-CONTAINING, MASS OF GAS . . .*
> OK
> *STAR?*
> A STAR IS A SELF-LUMINOUS, SELF-CONTAINING, MASS OF GAS . . .

---

*Italicized portion is the user input.

Fig. 6-28. Lombardo pak typical programming sequence.

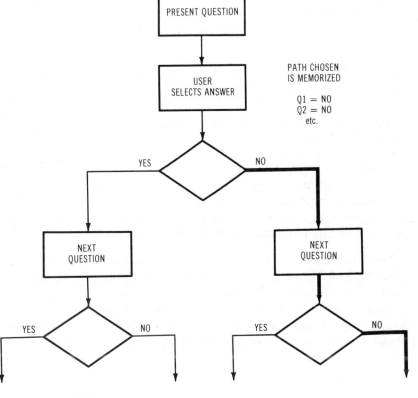

This dialog assumes you are teaching the robot general knowledge. In this particular application, you have stated a new word to the robot. This word, in this case "STAR," is then received by the operator prompt, "WHAT IS A STAR?" You would then type in the description of a star. Obviously, you cannot do this particular application with the hex keypad. This would have to be used with an optional ASCII keyboard. It is being shown here to illustrate some artificial-intelligence methods of communicating with robots. In fact, from now on, we will move away from the advanced core processor personal robot to a system that is not defined, hardware-wise.

Assume a computer system that has a large amount of memory or a large number of mass storage options built in that are automatically called when more memory is needed. Also assume that there is a voice output and ASCII keyboard input; there might even be some sort of CRT or alphanumeric display that will allow the operator to view what is input for errors. From here, let's look at some real robot-type applications utilizing this type of artificial intelligence programming or childlike learning capabilities. Suppose we have a robot vacuum cleaner. I'm sure all of you have wielded a vacuum cleaner at least once in your life. From this experience you might find an application such as this pleasing but in order to program a robot to clean a room using vacuum techniques may be problematic. The reasons being, that locations of certain objects change within a room. Chairs, tables, shoes, and other things that might be on a floor may move from day to day. Therefore, a robot that is able to teach itself a certain path, depending on what it sees in front of it, would be useful. This type of a teaching algorithm is shown in Fig. 6-29. In this figure, the use of an ultrasonic ranger or some type of collision or object detection system is utilized. I call this technique auto-adaptive because it is initially programmed by the user through the walk-through method, like the autoprogramming mode. From there, the program has the ability to change itself, depending on what it sees. Let's go over that in detail.

The following shows the logic behind a dialog between the robot and user:

> *LEARN**

> WHAT ROOM IS THIS/
> *3 (LIVING ROOM)*
> START ME IN THE CORNER, DRIVE ME THROUGH THE SEQUENCE, THEN PUSH A.

It is assumed that the robot knows it is being programmed to vacuum a rug in a certain room. For simplicity, the room is certain corner of the room. From here, the user would enter into the auto-type mode and drive the robot through its paces. Its paces would be a back-and-forth motion with a slight side-to-side change to effect a scanning-type arrangement from

---
**Italicized portion is the user's input*

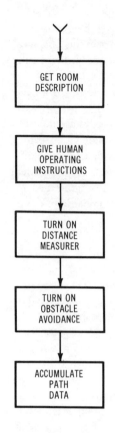

Fig. 6-29. Auto-adaptive technique in programming a robot vacuum cleaner.

one end of the room to the other. This type of back and forth, little to the left or right is well known by people who use vacuum cleaners but not so well known by robots. When it is done, you have successfully navigated the room manually. Whether you know it or not, you stopped in front of objects, you moved around chairs, you stopped the robot to pick up a pair of shoes, also, what you did not know is that, at this time, the robot was utilizing its sonic ranging system to get a readout of why you were stopping while you were doing it.

This type of transparent learning is not typically done to date. What would typically be done is that the program would learn all of these stops, starts, forward, and reverse motions without any regard for the reasons why. In an auto-adaptive system, the robot would participate in the learning but also exist at a higher level to try and find out what you were doing and why. By doing this, it is remembering when to stop or for what reasons to stop.

Fig. 6-30 is a representation of a robot vacuum cleaner moving through a room and encountering a chair. It is assumed that this figure is showing the learning mode being manually driven by an operator. Notice on the side of the figure there are certain areas in memory that are being messaged as the robot is driven through the room. These areas include an obstacle register, a direction register, and a status register. The obstacle is showing the feet to the next detected object. The direction is showing the current direction. There is also a direction change bit which will be consulted and further on down the figure is a memory status register. What

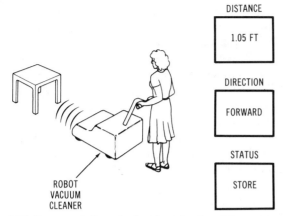

Fig. 6-30. Representation of robot learning lay of the room.

stopped the robot. The robot then, on that stop motion, consults its object register, direction change register and determines from this information that the reason it has been stopped is because of the obstacle that lies 1 foot in front of it. It then would present the memory status byte to indicate a remember (STORE) condition. Fig. 6-31 now shows a series of windows showing a robot maneuvering itself through the room after the learning has been accomplished. You will notice that it has remembered that when an obstacle is 1 foot in front of it, it is to stop and change direction. It is this type of adaptive learning that may be called artificial intelligence and should be employed in personal robots.

From here, let's move back to another area of robotics. One that we looked at several chapters ago, simulators. In simulator systems you try to be as realistic as possible. However, you can do some things with simulators that you cannot, presently, do with real robots. This may be due to mechanical or electrical limitations. In those conditions it might be possible to speculate on what a robot might be able to do. You can learn a lot through building one of these simulators. Let's take the XY table simulator that was discussed in Chapter 4. If you remember correctly, there were a series of very basic commands that allowed the operator to build a program to move blocks around an 8×8 grid. These blocks would be picked up by the XY table arm. In fact, I even showed some pictures of an actual implementation of the arm using erector set parts and stepper motors. Therefore, in the condition that we are going to talk about, it is possible to be able to implement it in real hardware. What I'd like to suggest is a mode of programming that I call "inherent." In the inherent mode the verbs: find, get, go, put, stop, do, and move are the action words that allow you to specify, in a sentence, what you want the table to do. For instance, instead of saying "GET 4.4" you might say "GET a large block" or "GET a black block." A statement might be made that might go like the following, "FIND a white block and put it on location 4.4." Obviously, this is a much higher-level command than the GET and PLACE commands were before. In fact, now you're requiring that the XY table go out and do the action of actually finding where a white block exists. But in this simple statement, "GET a large block" the computer that is controlling the table will see the inherent verb, get, ignore articles, and grasp the adjective noun phrase. It will then search its memory for something like it. For instance, if it knows "block" but not "large block" it will prompt the operator with the question, "What is large?" All answers are recorded for future reference. After a training session, soon the robot will be able to manipulate objects with the agility of a small child. Each time a new object is talked about, it is remembered.

You might recall that this programming like a small child has come up over the last few paragraphs quite a bit. That is the way to program robots. Inherent mode operation is a structured approach to this type of learning. See how many other high-level structures you can create for the XY table. Another exercise might be to create your own high-level simulator of an operator interface. You can do this through the use of a personal computer. Hang on a text-to-speech synthesizer and allow the operator to communicate with the computer on a free-flowing basis. While doing this communication, store all the results of logic operations for future use. I am currently doing research in providing programs such as the one I just examined. In a possible future book, that will be the subject of its contents.

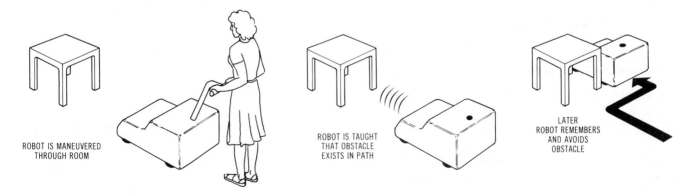

Fig. 6-31. Series of drawings showing robot navigating itself through a room consulting its "learned" experiences.

CONCLUSION

As I stated at the beginning of this chapter, this is the final unit in this book. It's the final unit of the first two books in this series. In the beginning, we covered rovers and all of their applications. If you'll notice in the past two chapters, many of the things we learned in the rover applications were directly applicable to personal robots. In fact, the manipulators system that we went over in Chapters 3 and 4 were also brought into play in these last two chapters. Personal robots are the future of the robotics business. Industrial pick-and-place and manufacturing systems will also grow. At the present time there is very little textual information on flexible manufacturing systems utilizing robots. As far as manipulators go, there is a wealth of knowledge available to date. A final word on personal robotics. Most of my work in robotics has been geared toward the personal robot.

I started writing for *Robotics Age Magazine* two years ago. It is these series of articles that have spawned the past three books. The introduction of HERO and the RB5X robot as well as the TOPO and BOB robots has served to spur my interest, and I hope yours also. Personal robots will undoubtedly grow in the future. In the next few years major manufacturers will show their wares, and there will be a few more robotics hobbyists added to the ranks. It is my hope that you will contribute to the field through continuing experimentation as I will. I hope to report the results of experimentation in volumes such as these.

I welcome any and all ideas, criticisms, and general friendly dialog. You can reach me through the mail at the address shown. I appreciate your comments. I cannot guarantee that I can get back to you on a timely basis. My regular feature articles in *Microcomputing* magazine net me a tremendous amount of mail. It is almost impossible to personally answer all of the technical questions that have been asked; however, you will get a reply, just have patience.

Mark J. Robillard
MJR Digital (ARS)
Box 630
Townsend, MA 01469

Index

A

Advanced
  CPU core, block diagram, HERO 1, 163-164
  hardware, HERO 1, 163-179
  robot language, 199-204
  software, preliminary EVENT-mode structure, 202
Allophone code chart, 172-173
All-ports message structure, 70, 76
AND port message structure, 70, 74
Arm-drive circuit board, HERO 1, block diagram, 161-162
Artificial intelligence, 204-208
Assembly, final, workstation design, 86-100
ATHOME command, CY500, 105, 106, 108
Auto-adaptive technique, programming, 206
Automated guideway vehicle, design specifications, 12-13

B

BASIC
  personal robots, 181-182
  RB5X robot, 181-182
  statements, 182
BITSET command, CY500, 106-107
Blank-screen routine, 128
Block diagram
  CDP1805, 41, 42
  CY500, 95, 98
  HERO 1, 150, 151-154
    advanced CPU core, 163-164
    arm-drive circuit board, 161-162
    CPU card, 151-152
    display, keyboard, 154-155
    intelligent display IC, 174, 179
    internal structure, 163-164
    I/O card, 152, 154
    light/sound-sensing device, 169-170, 172
    main drive board, 161-163
    motion detector, 160-161
    motor-controller subsystem, 172-173, 176
    operator-interface subsystem, 173
    power supply, 151
    robot interpreter, 193-194, 196
    sense circuit card, 160-161
    sonar receiver, 159-160
    sonar transmitter, 158-159
    sonic distance measurement system, 168-169
    speech accessory, 160
    system software architecture, 193, 196
    voice output subsystem, 170-171, 174
  MM54240, 27, 30
  M2, 40, 41
    motor control, 43-44
  receiver, MC14469, 21-22
  SCU20, 30, 33
  software, core processor, 200-201
  transmitter, MC14469, 24-25
  UCN4202A, 90
Bumper, remote feedback system, 39, 40
Buried wire guidance, 18-19

C

CD4052 dual four-channel analog gate, 38-39
CDP1805
  microprocessor, 41-44
  pinout, 41, 42
Character set definition, TI/994A, 127
Circuit description, HERO 1, 148-158
Circular
  coordinate system, 102
  robot manipulators, workstations, 87-88
CLEAR
  -BIT
    command, CY500, 106, 107
    message structure, 69, 74
  -event-counter message structure, 75, 80
Clock
  generation, 47-48
  oscillator configurations
    MC14469, 29, 32
    MM54240, 29, 32
Code chart allophone, 172-173
Coil mounting, 19
Color set definition, TI/994A, 127
Command
  -interpreter routine, 131
  language
    design, 110-121
    XY table, 103-104
    XY table, specification, 104
  message format, SCU20, 65, 72
  structure, CY500, 105-121
Commands, 103-104
  CY500
    ATHOME, 105, 106, 108
    BITSET, 106-107
    CLEARBIT, 106, 107
    DOIT, 106, 107
    DOITNOW, 108
    EDIT, 110
    ENTER, 106, 107
    EXECUTE, 110, 111
    FACTOR, 106, 109
    GO, 107
    HALFSTEP, 107
    INITIALIZE, 107
    JOG, 107-108
    LEFT/RIGHT, 108
    LOOPTIL, 108
    MINUS, 109
    NUMBER, 106, 108
    ONESTEP, 108
    PLUS, 109

Commands—cont
  CY500
    POSITION, 106, 107, 108
    QUIT, 108
    RATE, 106, 109
    SLOPE, 109
    WAITUNTIL, 109
    ZERO, 109
  HERO
    robot, 194-196
    SPEECH, 196-197
  I/O, 68-71, 72, 73, 74, 75, 76, 77, 78, 79
  timer, SCU20, 73-75
COMPARE/ROTATE ROBOL statements, 187-188, 191
Configuration, CY500, 96, 99
Connection, master, MC14469, 23-24
Control
  distributed, 20-35
  electronics, 19-35
    distributed, 20-35
    hard-wired single-function, 19
    programmable, 19-20
  node, motor, 25-26
  planning, 48-50
  programmable, 19-20
  speed, 55, 57-58, 60, 61, 62, 63, 64
Coordinate
  controller, listing, 122-124
  system
    circular, 102
    XY, 102-103
Core processor
  memory map, 165-168
  software, block diagram, 200-201
CPU
  card, HERO, 151-152
  core cards schematic, HERO 1, 163-165, 166
  interrupt circuit, HERO 1, schematic diagram, 152-153
CURSOR routine, 136
CY500, 92-100
  block diagram, 95, 98
  command structure, 105-121
  configuration, 96, 99
  interface, stepper motor, 94, 98
  keyboard, input hookup, 94, 97
  pinout, 97, 99
  schematic XY table interface, 98, 100
  serial port hookup, 94, 97

D

Data
  logging, 35, 75
  pak, 190-191, 194
  rate select code, SCU20, 31, 34
  structure, received, MC14469, 21-22
Debugger, HERO 1, 156-157
Design
  command language, 101-121
  motion system, 11-36
  noise immune communication links, MC14469, 23-24
  specifications
    automated guideway vehicle, 12-13
    mail-handling robot, 12-13
    security robot, 12-13
  workstation, final assembly, 86-100

Distance counter, MC14469, 26
Distributed
  control, 20-35
    MC14469, 21-27
    software, 65-66, 68-75, 76, 77, 78, 79, 80, 81, 189-192
      data pak, 190-191, 194
      EPROM, 191
      ROM, 191
      semiconductor memories, 190-191
DOIT command, CY500, 106, 107
DOITNOW command, CY500, 108
Draw
  -blocks-grid routine, 128
  -text-window routine, 129
Drive considerations, 13-14

E

EDIT
  command, CY500, 110
  routines, 114, 131-132
Eight-bit
  analog-to-digital converter, 25
  eight-channel converter, 26
1800 microprocessor, 40-42
Electrolyte, 35
Electronics, control, 19-35
ENTER command, CY500, 106, 107
EPROM, 191
Error-handling routines, 112-113
EXECUTE command, CY500, 110, 111
Executive routine, HERO, 193
EXTEND/RETRACT ROBOL statements, 186, 190

F

FACTOR command, CY500, 106, 109
Final assembly workstation design, 86-100
Flowcharts
  MC14469, 26, 28
  optical sensor, speed control, 57, 62
  speed controller, 55, 61
  symbols, 49
  two-detector system, 58, 64
FORTH, 183
FORWARD/REVERSE ROBOL statements, 188, 192
Four-wheeled drive mechanics, 13-14

G

GET
  -BLOCK routine, 134
  routine, 116-117
GO command, CY500, 107
GOTO routine, 116, 133
Guidance
  optical,
    one-head detector, 15-16
  rover, 14-19
  system
    buried wire, 18-19
    two-detector, 16-17
  track, 14-18

H

HALFSTEP command, CY500, 107
Handling material systems, 9-10
Hardware
    advanced, HERO 1, 163-179
    interface, wait logic, MM54240, 29, 32
    manipulator systems, 85-100
    robot, personal, 145-179
Hard-wired single-function machine, 19
HERO 1, 146-179
    advanced hardware, 163-179
    arm, wrist, gripper, 149-150
    block diagram, 150, 151-154
        advanced CPU core, 163-164
        arm-drive circuit board, 161-162
        CPU card, 151-152
        display, keyboard, 154-155
        intelligent display IC, 174, 179
        internal structure, 163-164
        light/sound sensing device, 169-170, 172
        main drive board, 161-163
        motion detector, 160-161
        motor-controller subsystem, 172-173, 176
        operator-interface subsystem, 173, 177
        robot interpreter, 193-194, 196
        sense circuit card, 160-161
        sonar receiver, 159-160
        sonar transmitter, 158-159
        sonic distance measurement system, 168-169
        speech accessory, 160
        system software architecture, 193, 196
        voice output subsystem, 170-171, 174
    body, 150
    circuit description, 148-158
    commands, SPEECH, 196-197
    core processor memory map, 165-168
    CPU card, 151-152
        interrupt circuit schematic, 152-153
    debugger, 156-157
    drive mechanism, 150
    Executive
        routine, 193
        system, 192-204
    head, 149
    interrupt code chart, 165
    I/O card, block diagram, 152, 154
    keyboard, 156
    memory locations, 165-166
    motor MOVE commands, 194-195
    phoneme codes, Votrax SC01A, 196-197
    power
        supply, 150
        supply block diagram, 151
        system, 150
    processor logic, 150-154
    program examples, 197-198
    robot commands, 194-196
    schematic
        CPU core card, 163-165, 166
        light/sound sensing device, 169-170, 174
        motion-sensing device, 169, 172
        sonic distance measurement system, 168-169, 170
    specifications, 149
    teaching pendant, 156-157, 198-199

Home
    personal robot requirements, 148
    routine, 118, 136

I

Industrial robot, 145
Initial input section, 111
Initialization
    routine, 111
    UCN4202A, 104-105
INITIALIZE command, CY500, 107
Input-text-handler routine, 129
Intelligence, artificial, 204-208
Intelligent display IC, HERO 1, block diagram, 174, 179
Interface memory locations, programmer's model, 120
Internal structure, HERO 1, block diagram, 163-164
Interrupt code chart, HERO 1, 165
I/O
    card, block diagram, HERO 1, 152, 154
    commands, 68-71, 72, 73, 74, 75, 76, 77, 78, 79

J

JOG command, CY500, 107-108

K

Keyboard, HERO 1, 156

L

Language
    design, command, 101-121
    robot, advanced, 199-204
Languages
    BASIC, 181-182
    FORTH, 183
    LOGO, 183
    ROBOL, 183-189
LEFT/RIGHT command, CY500, 108
LIFT/LOWER ROBOL statement, 186-187, 188
Light/sound sensing device, HERO 1,
    block diagram, 169-170, 172
    schematic, 169-170, 174
LIST routine, 115
LISTEN ROBOL statement, 185, 186
Load
    -all-ports message structure, 68-69, 74
    -port message structure, single, 68, 73
Logging data, 35, 75
Logical operators, 71-72
LOGO, 145-146, 183
Lombardo pak, 205
LOOPTIL command, CY500, 108

M

Mail
    -handling robot, design specifications, 12-13
    mobile software, 59-65, 66, 67, 68, 69, 70, 71, 72
    persons, 10
Main drive board, HERO 1, block diagram, 161-163
Manipulator
    robot, XY type design, 89-100
    software, 101-143

Manipulator—cont
  systems hardware, 85-100
    production environment, 85-86
Master-mode hookup, MM54240, 28, 31
Material handling systems, 9-10
MC14469
  block diagram, 21-22
    transmitter, 24-25
  clock oscillator configurations, 29, 32
  connection, 23-24
  data structure, received, 21-22
  design noise immune communication links, 23-24
  distance counter, 26
  8-bit
    a/d converter connection, 25
    8-channel converter, 26
  flowcharts, 26, 28
    pin designations, 22
  receiver/transmitter, 21-27
  rover system, 26-27
  timing diagram, 26, 29
Mechanical drive system two-motor, 51-52
Mechanics, moving platform, 9-46
Memories, semiconductor, 190-191
Memory locations, HERO 1, 165-166
Message structures
  all-ports, 70, 76
  AND port, 70, 74
  clear-bit, 69, 74
  clear-event-counter, 75, 80
  load-all-ports, 68-69, 74
  OR port, 70, 74
  read
    -event-counter, 75, 80
    -log-count, 75, 81
    ports, 70, 77
    -ports, mask-provided, 70-71, 78, 79
  set-bit, 69, 74
  single, load-port, 68-73
  start
    -event, 73-74, 80
    -log, 75, 81
  step-event-counter, 74-75, 80
  stop
    -and-read-log, 75, 81
    -event-counter, 75, 80
  test-bit, 70, 77
  toggle-bit, 69-70, 74
Microprocessor-interfaced
  optical guidepath, two-detector, 57, 63
  single-detector guidepath circuit, 57, 63
Microprocessors, 20
  CDP 1805, 41-44
  1800, 40-42
  SCU20, 30-35
  6808, 151-152
MINUS commands, CY500, 109
MM54240
  asynchronous receiver/transmitter, 27-30
  block diagram, 27, 30
  clock oscillator configurations, 29, 32
  hardware interface, wait logic, 29, 32
  master mode hookup, 28, 31
  output
    configuration input connections, 27, 30
    waveforms, 29, 32

MM54240—cont
  output
    pin configuration, 27, 29
M1
  remote, mobile platform, 37-39
  software considerations, 75-78, 80-81
Mostek SCU20 serial control unit, 30-35
Motion
  detector, HERO 1, block diagram, 160-161
  routine, 119-120
  -sensing device, HERO 1, schematic diagram, 169, 172
  system design, 11-36
Motor control
  MOVE commands, HERO, 194-195
  node, 25-26
  program, 53-55
  SCU20, 34, 35
  software, 50-52
Motor-controller subsystem, HERO 1, block diagram, 172-173, 176
Motorola, MC14469 addressable asynchronous receiver/transmitter, 21-27
MOVE routine, 117-118, 134-135
Moving platform mechanics, 9-46
M2
  block diagram, 40, 41
  intelligent mobile platform, 39-46
  motor control
    block diagram, 43-44
    schematic, 43, 45
  software description, 81-84
  ultrasonic ranger interface, 43-44, 46

N

National Semiconductor, MM54240, 27-30
Nicad cell, 36, 37
NUMBER commands, CY500, 106, 108

O

One
  -head optical guidance detector, 15-16
  STEP command, CY500, 108
OPEN/CLOSE ROBOL statements, 185-186, 187
Operator-interface subsystem, HERO 1, block diagram, 173, 177
Optical track guidance, 14-18
OR port message structure, 70, 74
Output
  configuration input connections, MM54240, 27, 30
  waveforms, MM54240, 29, 32

P

Pak,
  data, 190-191, 194
  Lombardo, 205
Parallel-to-serial shift register, 32, 34
Personal robot
  hardware, 145-179
  home requirements, 148
  software, 181-208
    BASIC, 181-189
  uses, 147
Phoneme codes, Votrax SC01A, HERO, 196-197
Pin
  configuration, MM54240, 27, 29

Pin—cont
  designations
    MC14469, 22
    SCU20, 31, 34
   -out, CDP1805, 41-42
   -out CY500, 97, 99
PITCH/YAW/ROLL ROBOL statements, 186, 189
PLACE
  -BLOCK routine, 135-136
  routine, 118
Planning, control, 48-50
Platform, moving, mechanics, 9-46
PLOT routine, 121, 136
PLUS command, CY500, 109
POSITION command, CY500, 106, 107, 108
Power
  supply, HERO 1, 150
  systems, 35-36, 37
Primary cells, 36
Processor logic, HERO 1, 150-154
Production environment, 85-86
Program
  examples, HERO, 197-198
  -input routine, 131
Programmable control, 19-20
  microprocessors, 20
Programmer's model
  interface memory locations, 120
  subprogram, 105
Programming auto-adaptive technique, 206

Q

QUIT command, CY500,108

R

RATE command, CY500, 106, 109
RB5X robot
  BASIC, 181-182
  specifications, 149
RCA CDP1805, 41-42
Read
  -event-counter message structure, 75, 80
  -log-count message structure, 75, 81
  -ports
    mask-provided message structure, 70-71, 78, 79
    message structures, 70, 77
Receiver/transmitter
  MC14469, 21-27
  MM54240, 27-30
Remote, mobile platform, M1, 37-39
Requirements, home personal robot, 148
Response message format, 66, 72
RIGHT/LEFT ROBOL statements, 188-189, 193
ROBOL, 183-189
  requirements, 183
  statements, 183-184
    EXTEND/RETRACT, 186, 190
    FORWARD/REVERSE, 188, 192
    LIFT/LOWER, 186-187, 188
    LISTEN, 185, 186
    OPEN/CLOSE, 185-186, 187
    PITCH/YAW/ROLL, 186, 189
    RIGHT/LEFT, 188-189, 193
    SAY, 185, 186

ROBOL—cont
  statements
    SCAN, 184
    SEE, 184-185
Robot
  hardware, personal, 145-179
  industrial, 145
  language, advanced, 199-204
  machine language, HERO, 197-198
  manipulator, XY type design, 89-100
  personal uses, 147
  security characteristics, 11
ROM, 191
  cartridge, 20
Routines
  blank-screen, 128
  command-interpreter, 131
  CURSOR, 136
  draw
    -blocks-grid, 128
    -text-window, 129
  EDIT, 114, 131-132
  error-handling, 112-113
  GET, 116-117
    -BLOCK, 134
  GOTO, 116, 133
  HOME, 118, 136
  initialization, 111
  input-text-handler, 129
  LIST, 115
  motion, 119-120
  MOVE, 117-118, 134-135
  PLACE, 118
    -BLOCK, 135-136
  PLOT, 121, 136
  program-input, 131
  RUN, 115-116, 133
  SCROLL, 129-130
  TARGET, 121, 136-137
    PLOT, 136
Rover
  guidance, 14-19
  system
    MC14469, 26-27
    SCU20, 72-73, 79
    software, 47-84
    software, clock generation, 47-48
Roving robot design, 11-36
RUN routine, 115-116, 133

S

SAY ROBOL statement, 185, 186
SCAN ROBOL statement, 184
Schematic
  HERO 1
    CPU core card, 163-165, 166
    light/sound sensing device, 169-170, 174
    motion-sensing device, 169, 172
    sonic distance measurement system, 168-169, 170
    XY table interface, CY500, 98, 100
SCROLL routine, 129-130
SCU20, 30-35
  block diagram, 30, 33
  command message format, 65, 72
  data rate select code, 31, 34

SCU20—cont
  motor control, 34, 35
  pin designations, 31, 34
  response message format, 66, 72
  rover system, 72-73, 79
  shift register, 32, 34
  timer commands, 73-75
  two-way half-duplex link, 31, 34
Secondary batteries, 36
Security
  robot
    characteristics, 11
    design specifications, 12-13
  systems, 10, 11, 12
SEE ROBOL statement, 184-185
Semiconductor memories, 190-191
Sense circuit card, HERO 1, block diagram, 160-161
Set-bit message structure, 69, 74
Simulation, XY table, 121-143
Single-function, hard-wired machine, 19
6808 microprocessor, 151-152
SLOPE command, CY500, 109
Software
  advanced, preliminary EVENT-mode structure, 202
  block diagram, core processor, 200-201
  distributed, 65-66, 68-75, 76, 77, 78, 79, 80, 81
  distribution, 189-192
    data pak, 190-191, 194
    EPROM, 191
    ROM, 191
    semiconductor memories, 190-191
  HERO system architecture, 193, 196
  mailmobile, 59-65, 66, 67, 68, 69, 70, 71, 72
  manipulators, 101-143
  M1, 75-78, 80-81
  motor control, 50-52
  M2, 81-84
  personal robots, 181-208
  rover systems, 47-84
Sonar HERO 1 block diagram
  receiver, 159-160
  transmitter, 158-159
Sonic distance measurement system, HERO 1
  block diagram, 168-169
  schematic diagram, 168-169, 170
Speech
  accessory, HERO 1, block diagram, 160
  commands, HERO, 196-197
Speed control, 55, 57-58, 60, 61, 62, 63, 64
  flowchart
    optical sensor, 57, 62
    two-detector system, 58, 64
  microprocessor-interfaced
    optical guidepath two-detector, 57, 63
    single-detector guidepath circuit, 57, 63
  tachometer, 55, 61
Speed controller, three-speed, voltage-controlled, 55, 61
  flowchart, 55, 61
Sprague UCN4202A, 89-92
Sprite, 126
Start-event-counter message structure, 73-74, 80
Start-log message structure, 75, 81
Statements
  BASIC, 182
  ROBOL, 183-184
Steering wheel, 52-55

Step-event-counter message structure, 74-75, 80
Stop
  -and-read-log message structure, 75, 81
  -event-counter message structure, 75, 80
Symbols, flowchart, 49
Systems
  material handling, 9-10
  motion design, 11-36
  power, 35-36, 37
  security, 10, 11, 12

T

Tachometer, 55, 61
TARGET
  PLOT routine, 136
  routine, 121, 136-137
  world, 137
Teaching pendant HERO 1, 156-157, 198-199
Terrapin turtle, 145
Test-bit message structures, 70, 77
Three
  -speed voltage-controlled speed controller, 55, 61
  -wheeled drive mechanics, 13-14
TI/994A
  character set definition, 127
  color set definition, 127
Timer commands, SCU20, 73-75
Timing diagram, MC14469, 26, 29
Toggle-bit message structure, 69-70, 74
Turtle, Terrapin, 145
Two
  -detector guideway system, 16-17
  -motor mechanical drive system, 51-52
  -way half-duplex link, SCU20, 31, 34
  -wheeled drive mechanics, 13

U

UCN4202A
  block diagram, 90
  initialization, 104-105
  Sprague, 89-92
Ultrasonic ranger interface, M2, 43-44, 46
Uses, personal robot, 147

V

VIP controller board, 1800, 42-43
Voice-output subsystem, HERO 1
  block diagram, 170-171, 174
  schematic diagram, 171-172, 175
Voltaic cell, 35-36
Votrax SC01A phoneme codes, HERO, 196-197

W

WAITUNTIL command, CY500, 109
Wheel, steering, 52-55
Wire-guidance-system operation, 18-19
Workstations
  circular robot manipulators, 87-88
  final assembly, design, 86-100

X

X-axis
  cart, 92, 93
  drive chain, 92, 93
XY
  coordinate system, 102-103
  table, 103-104
    interface UCN4202A stepper controllers, 92, 96
    specification, 104
  type robot manipulator
    design, 89-100
    workstation, 87-89

XY table
  simulation, 121-143
  simulator, 137-143

Y

Y-axis assembly, 92, 94

Z

Z-axis assembly, linear stepper motor, 92, 94
ZERO command, CY500, 109